Columba Sara Evelyn (Ed.)

Heinz Ebert

Columba Sara Evelyn (Ed.)

Heinz Ebert

Natural history, Geologist, Teacher, Chemistry, Physics

Fec Publishing

Publisher:
Fec Publishing is a trademark of
International Book Market Service Ltd., 17 Rue Meldrum, Beau Bassin, 1713-01 Mauritius
Email: info@bookmarketservice.com
Website: www.bookmarketservice.com

Published in 2011

Printed in: U.S.A., U.K., Germany. This book was not produced in Mauritius.

ISBN: 978-613-6-89699-1

Contents

Articles

References

Article Licenses

Heinz Ebert

Heinz Ebert (Saxony, Germany, October 13, 1907, – 1983, Rio Claro, São Paulo) was a German-Brazilian geologist, naturalist, and educator.

Ebert had his basic training in chemistry, physics, and analytical chemistry and natural sciences and served as geologist at the Geological Survey of Prussia. He was obtained his PhD at the University of Leipzig, where he became an associate professor in 1934. He worked on petrography at the Geological Survey of Saxony and developed methods to recognize stratigraphic sequences by petrographic and structural analysis. From 1939 to 1946, World War II interrupted his geological activity. After the war he returned to geology and opened an office for applied geology (groundwater, drilling, economic geology, geology for dams, etc.)

He moved to Recife in 1950 and was influential in creating the CAGE (Campaign for Training Geologists). From 1950 to 1956, he was a "specialized geologist" at the Division of Geology and Mineralogy, which evolved from the old Geological Survey of Brazil, and mapped the regions of São João del Rei, Juiz de Fora, and Barbacena, a whole complex metasedimentary southern region of Minas Gerais. He was director of Sudene, training technicians in areas of groundwater in crystalline rocks.

In 1962 he moved to Rio Claro, São Paulo, to teach various subjects in the field of geosciences at the Faculty of Philosophy, Sciences and Letters of the Universidade Estadual Paulista Júlio de Mesquita Filho, assisting in the creation of the geology course. He continued his studies with attention to the south and southwest of the state of Minas Gerais. He made several trips abroad, bringing to Brazil contributions of his observations and contacts. He had many students, many of them becoming professors and continuing his work, especially in teaching mineralogy, petrology, structural geology, and the creation and maintenance of the Museum of Minerals and Rocks "Heinz Ebert".[1] This museum has a vast collection that Ebert donated, including many didactic collections.

Ebert formed in Recife more graduate students. He published around 60 works on geosciences, especially of Brazil. Between 1977 and 1980 he also collaborated with the course of geology at the Universidade Federal de Mato Grosso.

In 1974, Ebert won the Jose Bonifacio Gold Medal of the Sociedade Brasileira de Geologia,[2] the highest prize in Brazilian geology.

As a naturalist, Ebert had a passion for butterfly collection. He died in 1983, in Rio Claro, Brazil.

See also

List of geologists

References

[1] http://www.rc.unesp.br/museudpm/ebert.html (pt) about Dr. Heinz Ebert

[2] Heinz Ebert - José Bonifácio Gold Medal, 1974 (http://www.sbgeo.org.br/index.php/premiados/listar/2)

External links

- About Heinz Ebert (http://www.rc.unesp.br/museudpm/ebert.html)
- Museum of Rocks and Minerals "Heinz Ebert" at [[Rio Claro, São Paulo (http://www.rc.unesp.br/museudpm/entrar.html)]]

Natural history

Natural history is the scientific research of plants or animals, leaning more towards observational rather than experimental methods of study, and encompasses more research published in magazines than in academic journals.[1] Grouped among the natural sciences, natural history is the systematic study of any category of natural objects or organisms.[2] That is a very broad designation in a world filled with many narrowly focused disciplines. So while modern natural history dates historically from studies in the ancient Greco-Roman world and the medieval Arabic world through to the scattered European Renaissance scientists working in near isolation, today's field is more of a cross discipline umbrella of many specialty sciences. For example, geobiology has a strong multi-disciplinary nature combining scientists and scientific knowledge of many specialty sciences.

A person who studies natural history is known as a **naturalist** or "natural historian". Natural history is categorized among the natural sciences. As a published topic, it originated from studies in the ancient Greco-Roman world. The modern topic comprises many specialty sciences such as geobiology.

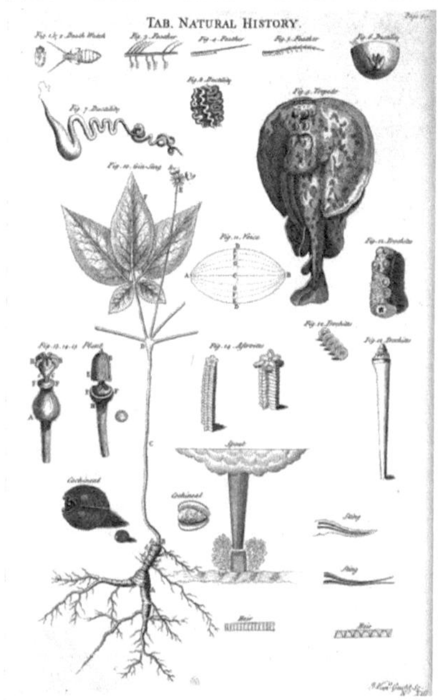

Tables of natural history, from the 1728 *Cyclopaedia*

Definitions

Historical

The English term 'natural history' is a translation of the Latin *naturalis historia.* Its meaning has narrowed considerably over time (see also History below). In antiquity, it covered more-or-less anything which is connected with nature or which uses materials drawn from nature; see for example the contents of Pliny's encyclopedia of this title, published circa AD 77-79.

Until well into the nineteenth century, knowledge was considered by Europeans to have two main divisions: the humanities (including theology), and studies of nature. Studies of nature could in turn be divided, with natural history being the descriptive counterpart to natural philosophy which was the analytical study of nature. In modern terms, natural philosophy roughly corresponded to modern physics and chemistry, while natural history included the biological and geological sciences. The two were strongly associated. During the heyday of the gentleman scientists, many figures contributed to both fields, and early papers in both were commonly read at professional science society meetings such as the Royal Society and the French Academy of Sciences – both founded during the seventeenth century.

Modern

Modern definitions of natural history come from a variety of fields and sources, and many of the modern definitions emphasize a particular aspect of the field, creating a plurality of definitions with a number of common themes among them. For example, while natural history if most often defined as a type of observation and a subject of study, it can also be defined as a body of knowledge, and as a craft or a practice, in which the emphasis is placed more on the observer than on the observed.[3]

Modern definitions from biologists often focus on the scientific study of individual organisms in their environment, as seen in this definition by M. Bates: "Natural history is the study of animals and Plants - of organisms. ... I like to think, then, of natural history as the study of life at the level of the individual - of what plants and animals do, how they react to each other and their environment, how they are organized into larger groupings like populations and communities"[4] and this more recent definition by D.S. Wilcove and T. Eisner: "The close observation of organisms—their origins, their evolution, their behavior, and their relationships with other species".[5] This focus on organisms in their environment is also echoed by H.W. Greene and J.B. Losos: "Natural history focuses on where organisms are and what they do in their environment, including interactions with other organisms. It encompasses changes in internal states insofar as they pertain to what organisms do".[6] Some definitions go further, focusing on direct observation of organisms in their environment, both past and present, such as this one by G.A. Bartholomew: "A student of natural history, or a naturalist, studies the world by observing plants and animals directly. Because organisms are functionally inseparable from the environment in which they live and because their structure and function cannot be adequately interpreted without knowing some of their evolutionary history, the study of natural history embraces the study of fossils as well as physiographic and other aspects of the physical environment".[7] A common thread in many definitions of natural history is the inclusion of a descriptive component, as seen in a recent definition by H.W. Greene: "Descriptive ecology and ethology".[8]

Recently, several authors have argued for a more expansive view of natural history, including S. Herman, who defines the field as "the scientific study of plants and animals in their natural environments. It is concerned with levels of organization from the individual organism to the ecosystem, and stresses identification, life history, distribution, abundance, and inter-relationships. It often and appropriately includes an esthetic component",[9] and T. Fleischner, who defines the field even more broadly, as “A practice of intentional, focused attentiveness and receptivity to the more-than-human world, guided by honesty and accuracy".[10] These definitions explicitly include the arts in the field of natural history, and are aligned with the broad definition outlined by B. Lopez, who defines the field as the "Patient interrogation of a landscape" while referring to the natural history knowledge of the Eskimo.[11]

A slightly different, but equally expansive framework for natural history is also implied in the scope of work encompassed by many leading natural history museums, which often include elements of Anthropology, Geology, Paleontology and Astronomy along with Botany and Zoology,[12] [13] or include both cultural and natural components of the world.[14] [15]

The plurality of definitions for this field has recently been recognized as both a weakness and a strength,[16] and a range of definitions have recently been offered by practitioners in a recent collection of views on natural history.[17]

History

Natural history begins with Aristotle and other ancient philosophers who analyzed the diversity of the natural world. Natural history, as a discipline, had existed since classical times, and fifteenth-century Europeans were very familiar with Pliny the Elder's *Historia Naturalis*. From the ancient Greeks until the work of Carolus Linnaeus (also known as Carl Linnaeus, or Carl von Linné) and other 18th century naturalists, the main concept of natural history was the *scala naturae* or Great Chain of Being, a conceptual arrangement of minerals, vegetables, more primitive forms of animals, and more complex life forms on a linear scale of increasing "perfection", culminating in our species.

Dioscorides' *De Materia Medica* is often said to be the oldest and most valuable work in the history of botany.[18] A Greek manuscript of Aristotle's *Biological Works*, written in Constantinople in the mid-9th century, and preserved at Corpus Christi College, Oxford, is probably the oldest surviving manuscript of texts that founded the science of biology.[19]

While natural history was basically static in medieval Europe, it continued to be developed by Arabic scholars during the Arab Agricultural Revolution. Al-Jahiz described early natural history ideas such as the "struggle for existence" (Malthus' phrase),[20] and the idea of a food chain.[21] He was an early adherent of environmental determinism.[22] Al-Dinawari is considered the founder of Arabic botany for his *Book of Plants*, in which he described at least 637 plants and discussed plant development from germination (sprouting) to death, describing the phases of plant growth and the production of flowers and fruit.[23] Abu al-Abbas al-Nabati developed an early scientific method for botany, introducing empirical and experimental techniques in the testing, description and identification of numerous materia medica, and separating unverified reports from those supported by actual tests and observations.[24] His student Ibn al-Baitar wrote a pharmaceutical encyclopedia describing 1,400 plants, foods, and drugs, 300 of which were his own original discoveries. A Latin translation of his work was useful to European biologists and pharmacists in the 18th and 19th centuries.[25] Earth sciences such as geology were also studied extensively by Arabic geologists, but by Avicenna's time, around 1000, the Arab Empire was in decline and scientists were not free to publish their ideas.[26]

From the 13th century, the work of Aristotle was adapted rather rigidly into Christian philosophy, particularly by Thomas Aquinas, forming the basis for natural theology. During the Renaissance, scholars (herbalists and humanists, particularly) returned to direct observation of plants and animals for natural history, and many began to accumulate large collections of exotic specimens and unusual monsters. Andrea Cesalpino was the creator of one of the first herbaria and the inventor of botanical systematics. Leonhart Fuchs was one of the three founding fathers of botany, along with Otto Brunfels and Hieronymus Bock. Important contributors to the field were also Valerius Cordus, Konrad Gesner (*Historiae animalium*), Frederik Ruysch, or Gaspard Bauhin.[19] The rapid increase in the number of known organisms prompted many attempts at classifying and organizing species into taxonomic groups, culminating in the system of the Swedish naturalist Carl Linnaeus.[19]

Georges Buffon is best remembered for his *Histoire naturelle*, a 44 volume encyclopedia describing everything known about the natural world.

In modern Europe, professional disciplines such as physiology, botany, zoology, geology, and palaeontology were formed. *Natural history*, formerly the main subject taught by college science professors, was increasingly scorned by scientists of a more specialized manner and relegated to an "amateur" activity, rather than a part of science proper. In Victorian Scotland it was believed that the study of natural history contributed to good mental health.[27] Particularly in Britain and the United States, this grew into specialist hobbies such as the study of birds, butterflies, seashells (malacology/conchology), beetles and wildflowers; meanwhile, scientists tried to define a unified discipline of biology (though with only partial success, at least until the modern evolutionary synthesis). Still, the traditions of natural history continue to play a part in the study of biology, especially ecology (the study of natural systems involving living organisms and the inorganic components of the Earth's biosphere that support them), ethology (the scientific study of animal behavior), and evolutionary biology (the study of the relationships between life-forms over very long periods of time), and re-emerges today as integrative organismal biology.

Amateur collectors and natural history entrepreneurs played an important role in building the large natural history collections of the nineteenth and early twentieth centuries, such as the Smithsonian Institution's National Museum of Natural History.

Museums

Natural history museums, which evolved from cabinets of curiosities, played an important role in the emergence of professional biological disciplines and research programs. Particularly in the 19th century, scientists began to use their natural history collections as teaching tools for advanced students and the basis for their own morphological research.

Societies

The term "natural history" alone, or sometimes together with archeology, forms the name of many national, regional and local natural history societies that maintain records for birds (ornithology), mammals (mammalogy), insects (entomology), fungi (mycology) and plants (botany). They may also have microscopical and geological sections.

Examples of these societies in Britain include the Natural History Society of Northumbria founded in 1829, British Entomological and Natural History Society founded in 1872, Birmingham Natural History Society, Glasgow Natural History Society, London Natural History Society founded in 1858, Manchester Microscopical and Natural History Society established in 1880, Scarborough Field Naturalists' Society and the Sorby Natural History Society, Sheffield, founded in 1918. The growth of natural history societies was also spurred due to the growth of British colonies in tropical regions with numerous new species to be discovered. Many civil servants took an interest in their new surroundings, sending specimens back to museums in Britain. (See also Indian natural history)

See also

- Big History
- Evolutionary history of life
- Natural environment
- Natural landscape
- Natural philosophy
- Natural science
- Naturalism (philosophy)
- Nature documentary
- Nature writing
- Nature
- Nature study
- Prehistoric life
- Terra: The Nature of Our World (video podcast)
- Timeline of evolution

- - - - -

- Parson-naturalist
- Russian naturalists

References

Citations and notes

[1] Natural History (http://wordnetweb.princeton.edu/perl/webwn?s=natural+history) WordNet Search, princeton.edu.

[2] Brown, Lesley (1993). *The New shorter Oxford English dictionary on historical principles*. Oxford [Eng.]: Clarendon. ISBN 0-19-861271-0.

[3] Thomas Lowe Fleischner, The Way of Natural History (Trinity University Press, 2011)

[4] Marston Bates, The nature of natural history (New York: Scribners, 1954)

[5] D. S Wilcove and T. Eisner, "The impending extinction of natural history," Chronicle of Higher Education 15 (2000): B24

[6] H. W. Greene and J. B. Losos, "Systematics, Natural-History, and Conservation - Field Biologists Must Fight a Public-Image Problem," Bioscience 38 (1988): 458-462

[7] G. A. Bartholomew, "The Role of Natural History in Contemporary Biology", Bioscience 36 (1986): 324-329

[8] H.W. Greene, "Organisms in nature as a central focus for biology," Trends in Ecology & Evolution 20 (2005):23-27

[9] S. G Herman, "Wildlife biology and natural history: time for a reunion," The Journal of wildlife management 66, no. 4 (2002): 933–946

[10] T. L. Fleischner, "Natural history and the spiral of offering," Wild Earth 11, no. 3/4 (2002): 10–13

[11] Barry Lopez, Arctic Dreams (Vintage, 1986)

[12] Amerian Museum of Ntural History, Mission Statement, http://www.amnh.org/about/

[13] Field Museum, Mission Statement, http://fieldmuseum.org/about/mission

[14] The Natural History Museum, Mission Statement, http://www.nhm.ac.uk/about-us/index.html

[15] National Natural History Museum of Chile, Mission Statement, http://www.dibam.cl/historia_natural/contenido.asp?id_contenido=277&id_submenu=650&id_menu=43

[16] http://declinetorebirth.org/conversations/an-accepted-way-of-viewing-art

[17] http://declinetorebirth.org/

[18] Gulsel M. Kavalali (2003). " *Urtica: therapeutic and nutritional aspects of stinging nettles* (http://books.google.com/books?id=AoWtF1ruQJsC&pg=PA15&dq&hl=en#v=onepage&q=&f=false)". CRC Press. p.15. ISBN 041530833X

[19] " Natural History Timeline (http://www.historyofscience.com/G2I/timeline/index.php?category=Natural+History)". HistoryofScience.com.

[20] Conway Zirkle (1941), Natural Selection before the "Origin of Species", *Proceedings of the American Philosophical Society* **84** (1): 71-123.

[21] Frank N. Egerton, "A History of the Ecological Sciences, Part 6: Arabic Language Science - Origins and Zoological", *Bulletin of the Ecological Society of America*, April 2002: 142-146 [143]

[22] Lawrence I. Conrad (1982), "Taun and Waba: Conceptions of Plague and Pestilence in Early Islam", *Journal of the Economic and Social History of the Orient* **25** (3), pp. 268-307 [278].

[23] Fahd, Toufic, "Botany and agriculture", pp. 815, in Morelon, Régis; Rashed, Roshdi (1996), *Encyclopedia of the History of Arabic Science*, **3**, Routledge, ISBN 0415124107

[24] Huff, Toby (2003), *The Rise of Early Modern Science: Islam, China, and the West*, Cambridge University Press, p. 218, ISBN 0521529948

[25] Diane Boulanger (2002), "The Islamic Contribution to Science, Mathematics and Technology", *OISE Papers*, in *STSE Education*, Vol. 3.

[26] Richard Myers (2003). " *The Basics of Chemistry* (http://books.google.com/books?id=oS50J3-IfZsC&pg=PA13&dq&hl=en#v=onepage&q=&f=false)". Greenwood Publishing Group. p.13. ISBN 0313316643

[27] Diarmid A. Finnegan (2008), "'An aid to mental health': natural history, alienists and therapeutics in Victorian Scotland", *Studies in History and Philosophy of Biological and Biomedical Sciences* **39** (3): 326–337, doi:10.1016/j.shpsc.2008.06.006, PMID 18761284

General information

- Allen, David Elliston (1994), *The Naturalist in Britain: a social history*, New Jersey: Princeton University Press, pp. 270, ISBN 0-691-03632-2
- Peter Anstey, " Two Forms of Natural History (http://blogs.otago.ac.nz/emxphi/2011/01/two-forms-of-natural-history/)", *Early Modern Experimental Philosophy* (http://blogs.otago.ac.nz/emxphi/), 17 January 2011.
- Atran, Scott (1990), *Cognitive Foundations of Natural History: Towards an Anthropology of Science*, Cambridge, UK: Cambridge University Press, pp. 376, ISBN 978-0521438711
- Kohler, Robert E. *Landscapes and Labscapes: Exploring the Lab-Field Border in Biology*. University of Chicago Press: Chicago, 2002.
- Mayr, Ernst. *The Growth of Biological Thought: Diversity, Evolution, and Inheritance*. The Belknap Press of Harvard University Press: Cambridge, Massachusetts, 1982.
- Rainger, Ronald; Keith R. Benson; and Jane Maienschein, editors. *The American Development of Biology*. University of Pennsylvania Press: Philadelphia, 1988.

External links

- A History of the Ecological Sciences by Frank N. Egerton (http://esapubs.org/bulletin/current/history_links_list.htm)

Geologist

A **geologist** is a scientist who studies the solid and liquid matter that constitutes the Earth as well as the processes and history that has shaped it. Geologists usually engage in studying geology. Geologists, studying more of an applied science than a theoretical one, must approach Geology using physics, chemistry and biology as well as other sciences. Geologists, compared to scientists engaged in other fields, are generally more exposed to the outdoors than staying in laboratories; although some geologists prefer to perform most of their studies in the lab.

The geologist - Carl Spitzweg, circa 1860

Geologists are engaged in exploration for mining companies in search of metals, oils, and other Earth resources. They are also in the forefront of natural hazards and disasters warning and mitigation, studying earthquakes, volcanic activity, tsunamis, weather storms, and the like; their studies are used to warn the general public of the occurrence of these events. Currently, geologists are also the scientists most engaged in the discussion of climate change, being the ones who hold the history and evidence for this Earth process.

Geologists exploring Jurassic sedimentary rocks in Makhtesh Gadol, Negev Desert, Israel

Geologists explaining the importance of volcanic ash layers to students on field in Iceland

Training / Schooling

Students at geology field school, University of Zimbabwe

Their training typically includes significant coursework in physics, mathematics, and chemistry, in addition to classes offered through the geology department; historical and physical geology, igneous and metamorphic petrology and petrography, hydrogeology, sedimentology, stratigraphy, mineralogy, palaeontology, physical geography and structural geology are among the many required areas of study. Most geologists also need skills in GIS and other mapping techniques. Geology students often spend portions of the year, especially the summer though sometimes during a January term, living and working under field conditions with faculty members (often referred to as "field camp"). Many non-geologists often take geology courses or have expertise in geology that they find valuable to their fields; this is common in the fields of geography, engineering, chemistry, urban planning, environmental studies, among others.

Areas of specialization

Geologist mud logging, common in petroleum and water-well drilling

Geologists may concentrate their studies or research in one or more of the following disciplines:

- Economic geology: the study of ore genesis, and the mechanisms of ore creation, geostatistics.
- Engineering geology: application of the geologic sciences to engineering practice for the purpose of assuring that the geologic factors affecting the location, design, construction, operation and maintenance of engineering works are recognized and adequately provided for;
- Geophysics: the applied branch deals with the application of physical methods such as gravity, seismicity, electricity, magnetic properties to study the earth.
- Geochemistry: the applied branch deals with the study of the chemical makeup and behaviour of rocks, and the study of the behaviour of their minerals.
- Geochronology: the study of isotope geology specifically toward determining the date within the past of rock formation, metamorphism, mineralization and geological events (notably, meteorite impacts).
- Geomorphology: the study of landforms and the processes that create them

- Hydrogeology: the study of the origin, occurrence and movement of groundwater water in a subsurface geological system.
- Igneous petrology: the study of igneous processes such as igneous differentiation, fractional crystallization, intrusive and volcanological phenomena .
- Isotope geology: the study of the isotopic composition of rocks to determine the processes of rock and planetary formation.
- Metamorphic petrology: the study of the effects of metamorphism on minerals and rocks.
- Marine geology: the study of the seafloor; involves geophysical, geochemical, sedimentological and paleontological investigations of the ocean floor and coastal margins. Marine geology has strong ties to physical oceanography and plate tectonics.
- Palaeoclimatology: the application of geological science to determine the climatic conditions present in the Earth's atmosphere within the Earth's history.
- Palaeontology: the classification and taxonomy of fossils within the geological record and the construction of a palaeontological history of the Earth.
- Pedology: the study of soil, soil formation, and regolith formation.
- Petroleum geology: the study of sedimentary basins applied to the search for hydrocarbons (oil exploration).
- Sedimentology: the study of sedimentary rocks, strata, formations, eustasy and the processes of modern day sedimentary and erosive systems.
- Structural geology: the study of folds, faults, foliation and rock microstructure to determine the deformational history of rocks and regions.
- Volcanology: the study of volcanoes, their eruptions, lavas, magma processes and hazards.

Employment

Professional geologists work for a wide range of government agencies, private firms, and non-profit and academic institutions. They are usually hired on a contract basis or hold permanent positions within private firms or official agencies (such as the USGS).

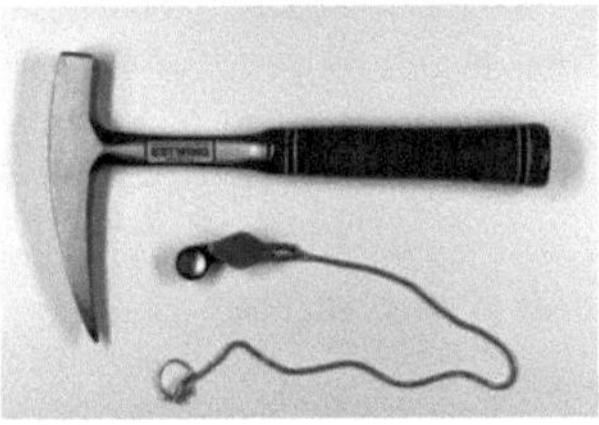

The rock hammer and hand lens (or loupe) are two of the most characteristic tools carried by geologists in the field.

Local, state, and national governments hire geologists to work on geological projects that are of interest to the public community. The investigation of a country's natural resources is often a key role when working for government institutions; the work of the geologist in this field can be made publicly available to help the community make more informed decisions related to the exploitation of resources, management of the environment and the safety of critical infrastructure - all of which is expected to bring greater wellbeing to the country.[1] This 'wellbeing' is often in the form of greater tax revenues from new or extended mining projects or through better infrastructure and/or natural disaster planning.

An engineering geologist is employed to investigate geologic hazards and geologic constraints for the planning, design and construction of public and private engineering projects, forensic and post-mortem studies, and environmental impact analysis. Exploration geologists use all aspects of geology and geophysics to locate and study natural resources. In many countries or US states without specialized environmental remediation licensure programs, such as Rhode Island and North Carolina, the environmental remediation field is often dominated by professional geologists, particularly hydrogeologists, with professional concentrations in this aspect of the field. Petroleum and mining companies use mudloggers and large-scale land developers use the skills of geologists and engineering geologists to help them locate oil and minerals, adapt to local features such as karst topography or earthquake risk, and comply with environmental regulations.

Geologists in academia usually hold an advanced degree in a specialized area within their geological discipline and are employed by universities.

See also

- List of geologists
- List of Russian geologists
- Petroleum geologist

"Picturesque camp made by a lone geologist on the cinders of Inferno," 1921

References

[1] Geoscience Australia, Our Role, December 16, 2010, http://www.ga.gov.au/about-us/our-role.html, Accessed: May 12, 2011

Teacher

Teacher

Classroom at a secondary school in Pendembu, Sierra Leone.

Occupation	
Names	Teacher, Educator, Lecturer
Type	Profession
Activity sectors	Education
Description	
Competencies	Teaching abilities, pleasant disposition, patience
Education required	Teaching certification
Fields of employment	Schools
Related jobs	Professor, academic, lecturer, tutor

Jewish children with their teacher in Samarkand, the beginning of the 20th century.

In education, a **teacher** (or, in the US, **educator**) is a person who provides schooling for pupils and students. A teacher who facilitates education for an individual student may also be described as a personal tutor. The role of teacher is often formal and ongoing, carried out by way of occupation or profession at a school or other place of formal education. In many countries, a person who wishes to become a teacher must first obtain professional qualifications or credentials from a university or college. These professional qualifications may include the study of pedagogy, the science of teaching. Teachers may have to continue their education after they qualify. Teachers may use a lesson plan to facilitate student learning, providing a course of study which is called the curriculum. A teacher's role may vary among cultures. Teachers may provide education instruction in literacy and numeracy, craftsmanship or vocational training, the Arts, religion or spirituality, civics, community roles, or life skills. In some countries, formal education can take place through home schooling.

Informal learning may be assisted by a teacher occupying a transient or ongoing role, such as a parent or sibling or within a family, or by anyone with knowledge or skills in the wider community setting.

Religious and spiritual teachers, such as gurus, mullahs, rabbis pastors/youth pastors and lamas may teach religious texts such as the Quran, Torah or Bible.

Professional educators

Teaching may be carried out informally, within the family which is called home schooling (see Homeschooling) or the wider community. Formal teaching may be carried out by paid professionals. Such professionals enjoy a status in some societies on a par with physicians, lawyers, engineers, and accountants (Chartered or CPA).

A teacher's professional duties may extend beyond formal teaching. Outside of the classroom teachers may accompany students on field trips, supervise study halls, help with the organization of school functions, and serve as supervisors for extracurricular activities. In some education systems, teachers may have responsibility for student discipline.

Around the world teachers are often required to obtain specialized education, knowledge, codes of ethics and internal monitoring.

There are a variety of bodies designed to instill, preserve and update the knowledge and professional standing of teachers. Around the world many governments operate teacher's colleges, which are generally established to serve and protect the public interest through certifying, governing and enforcing the standards of practice for the teaching profession.

The functions of the teacher's colleges may include setting out clear standards of practice, providing for the ongoing education of teachers, investigating complaints involving members, conducting hearings into allegations of professional misconduct and taking appropriate disciplinary action and accrediting teacher education programs. In many situations teachers in publicly funded schools must be members in good standing with the college, and private schools may also require their teachers to be college peoples. In other areas these roles may belong to the State Board of Education, the Superintendent of Public Instruction, the State Education Agency or other governmental bodies. In still other areas Teaching Unions may be responsible for some or all of these duties.

Pedagogy and teaching

In education, teachers facilitate student learning, often in a school or academy or perhaps in another environment such as outdoors. A teacher who teaches on an individual basis may be described as a tutor.

Dutch schoolmaster and children, 1662

A primary school teacher in northern Laos.

The teacher-student-monument in Rostock, Germany, honours teachers.

The objective is typically accomplished through either an informal or formal approach to learning, including a course of study and lesson plan that teaches skills, knowledge and/or thinking skills. Different ways to teach are often referred to as pedagogy. When deciding what teaching method to use teachers consider students' background knowledge, environment, and their learning goals as well as standardized curricula as determined by the relevant authority. Many times, teachers assist in learning outside of the classroom by accompanying students on field trips. The increasing use of technology, specifically the rise of the internet over the past decade, has begun to shape the way teachers approach their roles in the classroom.

GDR "village teacher" (a teacher teaching students of all age groups in one class) in 1951.

The objective is typically a course of study, lesson plan, or a practical skill. A teacher may follow standardized curricula as determined by the relevant authority. The teacher may interact with students of different ages, from infants to adults, students with different abilities and students with learning disabilities.

Teaching using pedagogy also involve assessing the educational levels of the students on particular skills. Understanding the pedagogy of the students in a classroom involves using differentiated instruction as well as supervision to meet the needs of all students in the classroom. Pedagogy can be thought of in two manners. First, teaching itself can be taught in many different ways, hence, using a pedagogy of teaching styles. Second, the pedagogy of the learners comes into play when a teacher assesses the pedagogic diversity of his/her students and differentiates for the individual students accordingly.

Perhaps the most significant difference between primary school and secondary school teaching is the relationship between teachers and children. In primary schools each class has a teacher who stays with them for most of the week and will teach them the whole curriculum. In secondary schools they will be taught by different subject specialists each session during the week and may have 10 or more different teachers. The relationship between children and their teachers tends to be closer in the primary school where they act as form tutor, specialist teacher and surrogate parent during the course of the day.

This is true throughout most of the United States as well. However, alternative approaches for primary education do exist. One of these, sometimes referred to as a "platoon" system, involves placing a group of students together in one class that moves from one specialist to another for every subject. The advantage here is that students learn from teachers who specialize in one subject and who tend to be more knowledgeable in that one area than a teacher who teaches many subjects. Students still derive a strong sense of security by staying with the same group of peers for all classes.

Co-teaching has also become a new trend amongst educational institutions. Co-teaching is defined as two or more teachers working harmoniously to fulfill the needs of every student in the classroom. Co-teaching focuses the student on learning by providing a social networking support that allows them to reach their full cognitive potential. Co-teachers work in sync with one another to create a climate of learning.

Rights to enforce school discipline

Throughout the history of education the most common form of school discipline was corporal punishment. While a child was in school, a teacher was expected to act as a substitute parent, with all the normal forms of parental discipline open to them.

In past times, corporal punishment (spanking or paddling or caning or strapping or birching the student in order to cause physical pain) was one of the most common forms of school discipline throughout much of the world. Most Western countries, and some others, have now banned it, but it remains lawful in the United States following a US Supreme Court decision in 1977 which held that paddling did not violate the US Constitution.[1]

Medieval schoolboy birched on the bare buttocks.

30 US states have banned corporal punishment, the others (mostly in the South) have not. It is still used to a significant (though declining) degree in some public schools in Alabama, Arkansas, Georgia, Louisiana, Mississippi, Oklahoma, Tennessee and Texas. Private schools in these and most other states may also use it. Corporal punishment in American schools is administered to the seat of the student's trousers or skirt with a specially made wooden paddle. This often used to take place in the classroom or hallway, but nowadays the punishment is usually given privately in the principal's office.

Official corporal punishment, often by caning, remains commonplace in schools in some Asian, African and Caribbean countries. For details of individual countries see School corporal punishment.

Currently detention is one of the most common punishments in schools in the United States, the UK, Ireland, Singapore and other countries. It requires the pupil to remain in school at a given time in the school day (such as lunch, recess or after school); or even to attend school on a non-school day, e.g. "Saturday detention" held at some US schools. During detention, students normally have to sit in a classroom and do work, write lines or a punishment essay, or sit quietly.

A modern example of school discipline in North America and Western Europe relies upon the idea of an assertive teacher who is prepared to impose their will upon a class. Positive reinforcement is balanced with immediate and fair punishment for misbehavior and firm, clear boundaries define what is appropriate and inappropriate behavior. Teachers are expected to respect their students, and sarcasm and attempts to humiliate pupils are seen as falling outside of what constitutes reasonable discipline.

Whilst this is the consensus viewpoint amongst the majority of academics, some teachers and parents advocate a more assertive and confrontational style of discipline. Such individuals claim that many problems with modern schooling stem from the weakness in school discipline and if teachers exercised firm control over the classroom they would be able to teach more efficiently. This viewpoint is supported by the educational attainment of countries—in East Asia for instance—that combine strict discipline with high standards of education.

It's not clear, however that this stereotypical view reflects the reality of East Asian classrooms or that the educational goals in these countries are commensurable with those in Western countries. In Japan, for example, although average attainment on standardized tests may exceed those in Western countries, classroom discipline and behavior is highly problematic. Although, officially, schools have extremely rigid codes of behavior, in practice many teachers find the students unmanageable and do not enforce discipline at all.

Where school class sizes are typically 40 to 50 students, maintaining order in the classroom can divert the teacher from instruction, leaving little opportunity for concentration and focus on what is being taught. In response, teachers may concentrate their attention on motivated students, ignoring attention-seeking and disruptive students. The result of this is that motivated students, facing demanding university entrance examinations, receive disproportionate resources. Given the emphasis on attainment of university places, administrators and governors may regard this policy as appropriate.

Obligation to honor students rights

Sudbury model democratic schools claim that popularly based authority can maintain order more effectively than dictatorial authority for governments and schools alike. They also claim that in these schools the preservation of public order is easier and more efficient than anywhere else. Primarily because rules and regulations are made by the community as a whole, thence the school atmosphere is one of persuasion and negotiation, rather than confrontation since there is no one to confront. Sudbury model democratic schools' proponents argue that a school that has good, clear laws, fairly and democratically passed by the entire school community, and a good judicial system for enforcing these laws, is a school in which community discipline prevails, and in which an increasingly sophisticated concept of law and order develops, against other schools today, where rules are arbitrary, authority is absolute, punishment is capricious, and due process of law is unknown.[2] [3]

Teacher enthusiasm

A teacher and young pupils at The British Museum Duveen Gallery

Since teachers can affect how students perceive the course materials, it has been found that teachers who showed enthusiasm towards the course materials and students can affect a positive learning experience towards the course materials. On teacher/course evaluations, it was found that teachers who have a positive disposition towards the course content tend to transfer their passion to receptive students.[4] These teachers do not teach by rote but attempt to find new invigoration for the course materials on a daily basis. ref>Motivated Student: Unlocking the Enthusiasm for Learning by Bob Sullo. Page 62</ref> One of the difficulties in this approach is that teachers may have repeatedly covered a curriculum until they begin to feel bored with the subject which in turn bores the students as well. Students who had enthusiastic teachers tend to rate them higher than teachers who didn't show much enthusiasm for the course materials.

Teachers that exhibit enthusiasm can lead to students who are more likely to be engaged, interested, energetic, and curious about learning the subject matter. Recent research has found a correlation between teacher enthusiasm and students' intrinsic motivation to learn and vitality in the classroom.[5] Controlled, experimental studies exploring intrinsic motivation of college students has shown that nonverbal expressions of enthusiasm, such as demonstrative gesturing, dramatic movements which are varied, and emotional facial expressions, result in college students reporting higher levels of intrinsic motivation to learn. Students who experienced a very enthusiastic teacher were more likely to read lecture material outside of the classroom.

There are various mechanisms by which teacher enthusiasm may facilitate higher levels of intrinsic motivation. Teacher enthusiasm may contribute to a classroom atmosphere full of energy and enthusiasm which feed student interest and excitement in learning the subject matter. Enthusiastic teachers may also lead to students becoming more self-determined in their own learning process. The concept of mere exposure indicates that the teacher's enthusiasm may contribute to the student's expectations about intrinsic motivation in the context of learning. Also, enthusiasm may act as a "motivational embellishment"; increasing a student's interest by the variety, novelty, and surprise of the enthusiastic teacher's presentation of the material. Finally, the concept of emotional contagion, may also apply. Students may become more intrinsically motivated by catching onto the enthusiasm and energy of the teacher.

Research shows that student motivation and attitudes towards school are closely linked to student-teacher relationships. Enthusiastic teachers are particularly good at creating beneficial relations with their students. Their ability to create effective learning environments that foster student achievement depends on the kind of relationship they build with their students.[6] [7] [8] [9] Useful teacher-to-student interactions are crucial in linking academic success with personal achievement.[10] Here, personal success is a student's internal goal of improving himself, whereas academic success includes the goals he receives from his superior. A teacher must guide his student in aligning his personal goals with his academic goals. Students who receive this positive influence show stronger self-confidence and greater personal and academic success than those without these teacher interactions.[9] [11] [12]

Students are likely to build stronger relations with teachers who are friendly and supportive and will show more interest in courses taught by these teachers.[10] [11] Teachers that spend more time interacting and working directly with students are perceived as supportive and effective teachers. Effective teachers have been shown to invite student participation and decision making, allow humor into their classroom, and demonstrate a willingness to play.[7]

The way a teacher promotes the course they are teaching, the more the student will get out of the subject matter. The three most important aspects of teacher enthusiasm are enthusiasm about teaching, enthusiasm about the students, and enthusiasm about the subject matter. A teacher must enjoy teaching. If they do not enjoy what they are doing, the students will be able to tell. They also must enjoy being around their students. A teacher who cares for their students is going to help that individual succeed in their life in the future. The teacher also needs to be enthusiastic about the subject matter they are teaching. For example, a teacher talking about chemsitry needs to enjoy the art of chemistry and show that to their students. A spark in the teacher may create a spark of excitement in the student as well. A enthusiastic teacher has the ability to be very influential in the young students life.

Misconduct

Misconduct by teachers, especially sexual misconduct, has been getting increased scrutiny from the media and the courts.[13] A study by the American Association of University Women reported that 9.6% of students in the United States claim to have received unwanted sexual attention from an adult associated with education; be they a volunteer, bus driver, teacher, administrator or other adult; sometime during their educational career.[14]

A study in England showed a 0.3% prevalence of sexual abuse by any professional, a group that included priests, religious leaders, and case workers as well as teachers.[15] It is important to note, however, that the British study referenced above is the only one of its kind and consisted of "a random ... probability sample of 2,869 young people between the ages of 18 and 24 in a computer-assisted study" and that the questions referred to "sexual abuse with a professional," not necessarily a teacher. It is therefore logical to conclude that information on the percentage of abuses by teachers in the United Kingdom is not explicitly available and therefore not necessarily reliable. The AAUW study, however, posed questions about fourteen types of sexual harassment and various degrees of frequency and included only abuses by teachers. "The sample was drawn from a list of 80,000 schools to create a stratified two-stage sample design of 2,065 8th to 11th grade students"Its reliability was gauged at 95% with a 4% margin of error.

In the United States especially, several high-profile cases such as Debra LaFave, Pamela Rogers, and Mary Kay Latourneau have caused increased scrutiny on teacher misconduct.

Chris Keates, the general secretary of National Association of Schoolmasters Union of Women Teachers, said that teachers who have sex with pupils over the age of consent should not be placed on the sex offenders register and that prosecution for statutory rape "is a real anomaly in the law that we are concerned about." This has led to outrage from child protection and parental rights groups.[16]

Teaching around the world

There are many similarities and differences among teachers around the world. In almost all countries teachers are educated in a university or college. Governments may require certification by a recognized body before they can teach in a school. In many countries, elementary school education certificate is earned after completion of high school. The high school student follows an education specialty track, obtain the prerequisite "student-teaching" time, and receive a special diploma to begin teaching after graduation.

International schools generally follow an English-speaking, Western curriculum and are aimed at expatriate communities.[17]

Canada

Teaching in Canada requires a post-secondary degree Bachelor's Degree. In most provinces a second Bachelor's Degree such as a Bachelor of Education is required to become a qualified teacher. Salary ranges from $40,000/year to $90,000/yr. Teachers have the option to teach for a public school which is funded by the provincial government or teaching in a private school which is funded by the private sector, businesses and sponsors.

England

Salaries for Nursery, Primary and Secondary School teachers ranged from £20,133 to £41,004 in September 2007, although some salaries can go much higher depending on experience and extra responsibilities.[18] Preschool teachers may earn £20,980 annually. Teachers in state schools must have at least a bachelor's degree, complete an approved teacher education program, and be licensed.

Many counties offer alternative licensing programs to attract people into teaching, especially for hard-to-fill positions. Excellent job opportunities are expected as retirements, especially among secondary school teachers, outweigh slowing enrollment growth; opportunities will vary by geographic area and subject taught.

France

In France, teachers, or *professors*, are mainly civil servants, recruited by competitive examination.

Republic of Ireland

Salaries for primary teachers in the Republic of Ireland depend mainly on seniority (i.e. holding the position of principal, deputy principal or assistant principal), experience and qualifications. Extra pay is also given for teaching through the Irish language, in a Gaeltacht area or on an island. The basic pay for a starting teacher is €30,904 p.a., rising incrementally to €59,359 for a teacher with 25 years' service. A principal of a large school with many years' experience and several qualifications (M.A., H.Dip., etc.) could earn over €90,000.[19]

Scotland

In Scotland, anyone wishing to teach must be registered with the General Teaching Council for Scotland (GTCS). Teaching in Scotland is an all graduate profession and the normal route for graduates wishing to teach is to complete a programme of Initial Teacher Education (ITE) at one of the seven Scottish Universities who offer these courses. Once successfully completed, "Provisional Registration" is given by the GTCS which is raised to "Full Registration" status after a year if there is sufficient evidence to show that the "Standard for Full Registration" has been met.[20]

For the salary year beginning April 2008, unpromoted teachers in Scotland earned from £20,427 for a Probationer, up to £32,583 after 6 years teaching, but could then go on to earn up to £39,942 as they complete the modules to earn Chartered Teacher Status (requiring at least 6 years at up to two modules per year.) Promotion to Principal Teacher positions attracts a salary of between £34,566 and £44,616; Deputy Head, and Head teachers earn from £40,290 to £78,642.[21]

United States

In the United States, each state determines the requirements for getting a license to teach in public schools. Teaching certification generally lasts three years, but teachers can receive certificates that last as long as ten years.[22] Public school teachers are required to have a bachelor's degree and the majority must be certified by the state in which they teach. Many charter schools do not require that their teachers be certified, provided they meet the standards to be highly qualified as set by No Child Left Behind. Additionally, the requirements for substitute/temporary teachers are generally not as rigorous as those for full-time professionals. The Bureau of Labor Statistics estimates that there are 1.4 million elementary school teachers,[23] 674,000 middle school teachers,[24] and 1 million secondary school teachers employed in the U.S.[25]

In the past, teachers have been paid relatively low salaries. However, average teacher salaries have improved rapidly in recent years. US teachers are generally paid on graduated scales, with income depending on experience. Teachers with more experience and higher education earn more than those with a standard bachelor's degree and certificate. Salaries vary greatly depending on state, relative cost of living, and grade taught. Salaries also vary within states where wealthy suburban school districts generally have higher salary schedules than other districts. The median salary for all primary and secondary teachers was $46,000 in 2004, with the average entry salary for a teacher with a bachelor's degree being an estimated $32,000. Median salaries for preschool teachers, however, were less than half the national median for secondary teachers, clock in at an estimated $21,000 in 2004.[26] For high school teachers, median salaries in 2007 ranged from $35,000 in South Dakota to $71,000 in New York, with a national median of $52,000.[27] Some contracts may include long-term disability insurance, life insurance, emergency/personal leave and investment options.[28] The American Federation of Teachers' teacher salary survey for the 2006-07 school year found that the average teacher salary was $51,009.[29] In a salary survey report for K-12 teachers, elementary school teachers had the lowest median salary earning $39,259. High school teachers had the highest median salary earning $41,855.[30] Many teachers take advantage of the opportunity to increase their income by supervising after-school programs and other extracurricular activities. In addition to monetary compensation, public school teachers may also enjoy greater benefits (like health insurance) compared to other occupations. Merit pay systems are on the rise for teachers, paying teachers extra money based on excellent classroom evaluations, high test scores and for high success at their overall school. Also, with the advent of the internet, many teachers are now selling their lesson plans to other teachers through the web in order to earn supplemental income, most notably on TeachersPayTeachers.com.[31]

Wales

Education in Wales differs in certain respects from education elsewhere in the United Kingdom. For example, a significant number of students all over Wales are educated either wholly or largely through the medium of Welsh: in 2008/09, 22 per cent of classes in maintained primary schools used Welsh as the sole or main medium of instruction. Welsh medium education is available to all age groups through nurseries, schools, colleges and universities and in adult education; lessons in the language itself are compulsory for all pupils until the age of 16.

Teachers in Wales can be registered emembers of trade unions such as NUT or NASUWT and reports in recent years suggest that the average age of teachers in Wales is falling with teachers being younger than in previous years.[32] A growing cause of concern are that attacks on teachers in Welsh schools which reached an all-time high between 2005 and 2010.[33]

Spiritual teacher

In Hinduism the spiritual teacher is known as a guru. In the Latter Day Saint movement the teacher is an office in the Aaronic priesthood, while in Tibetan Buddhism the teachers of Dharma in Tibet are most commonly called a Lama. A Lama who has through phowa and siddhi consciously determined to be reborn, often many times, in order to continue their Bodhisattva vow is called a Tulku.

There are many concepts of teachers in Islam, ranging from mullahs (the teachers at madrassas) to ulemas.

A Rabbi is generally regarded as the Jewish spiritual teacher.

Popular educators

- Howard Adelman
- Leib Glantz
- Charles Wedemeyer
- Edith Abbott
- Raymond Macdonald Alden
- Henry James Anderson
- Charles William Bardeen
- Charles Rollin
- Juan Pablo Bonet
- Lancelot Bavin
- Zarfishan Qaiser

See also

- Bullying in teaching
- Certified teacher
- College of Education
- Paraprofessional educator
- Student teacher
- Substitute teacher
- Teacher's union
- Teacher Support Network (in the UK)

References

[1] Ingraham v. Wright. (http://www.bucknell.edu/x4615.xml)

[2] The Crisis in American Education — An Analysis and a Proposal, The Sudbury Valley School (http://www.sudval.com/05_onepersononevote.html#02) (1970), *Law and Order: Foundations of Discipline* (http://books.google.com/books?id=MAqxzEss8k4C&pg=PA49&dq=The+Crisis+in+American+Education+â + An+Analysis+and+a+Proposal,+The+Sudbury+Valley+School+(1970),+Law+and+Order:+Foundations+of+Discipline) (pg. 49-55). Retrieved November 15, 2009.

[3] Greenberg, D. (1987) The Sudbury Valley School Experience "Back to Basics - Political basics." (http://www.sudval.com/05_underlyingideas.html#09)

> Knowing all this, we would expect; nay, insist (one would think) that the schools, in training their students to contribute productively to the political stability and growth of America, would be democratic and non-autocratic; be governed by clear rules and due process; be guardians of individual rights of students. A student growing up in schools having these features would be ready to move right into society at large. I think it is safe to say that the individual liberties so cherished by our ancestors and by each succeeding generation will never be really secure until our youth, throughout the crucial formative years of their minds and spirits, are nurtured in a school environment that embodies these basic American truths.

Retrieved January 4, 2010.

[4] Teaching Patterns: a Pattern Language for Improving the Quality of Instruction in Higher Education Settings by Daren Olson. Page 96

[5] Patrick, B.C., Hisley, J. & Kempler, T. (2000) "What's Everybody so Excited about?": The Effects of Teacher Enthusiasm on Student Intrinsic Motivation and Vitality", The Journal of Experimental Education, Vol. 68, No. 3, pp. 217-236

[6] Baker, J. A., Terry, T., Bridger, R., & Winsor, A. (1997). Schools as caring communities: A relational approach to school reform. School Psychology Review, 26, 576-588.

[7] Bryant, Jennings . 1980. Relationship between college teachers' use of humor in the classroom and students' evaluations of their teachers. Journal of educational psychology. 72, 4.

[8] Fraser, B. J., & Fisher, D. L. (1982). Predicting students' outcomes from their perceptions of classroom psychosocial environment. American Educational Research Journal, 19, 498- 518.

[9] Hartmut, J. (1978). Supportive dimensions of teacher behavior in relationship to pupil emotional cognitive processes. Psychologie in Erziehung und Unterricht, 25, 69-74.

[10] Osborne, E.;. Salzberger, I.; Wittenberg, G. W.. 1999. The Emotional Experience of Learning and Teaching. Karnac Books, London.

[11] Baker, J. A.Teacher-Student Interaction in Urban At-Risk Classrooms: Differential Behavior, Relationship Quality, and Student Satisfaction with School. The Elementary School Journal Volume 100, Number 1, 1999 by The University of Chicago.

[12] Moos, R. H. (1979). Evaluating Educational Environments: Measures, procedures, findings, and policy implications. San Francisco: Jossey-Bass.

[13] Goorian, Brad (December 1999). "Sexual Misconduct by School Employees." (http://web.archive.org/web/20080227085217/http://www.eric.ed.gov/ERICDocs/data/ericdocs2sql/content_storage_01/0000019b/80/15/fd/90.pdf) (PDF). *ERIC Digest* (134): 1. ERIC #: ED436816. Archived from the original (http://www.eric.ed.gov/ERICDocs/data/ericdocs2sql/content_storage_01/0000019b/80/15/fd/90.pdf) on 2008-02-27. . Retrieved 2008-01-17.

[14] Shakeshaft, Charol (June 2004). "Educator Sexual Misconduct: A Synthesis of Existing Literature" (http://www.ed.gov/rschstat/research/pubs/misconductreview/report.pdf#p28) (PDF). U.S. Department of Education, Office of the Under Secretary. pp. 28. . Retrieved 2008-01-17.

[15] Educator Sexual Misconduct: A Synthesis of Existing Literature (http://www.ed.gov/rschstat/research/pubs/misconductreview/report.pdf#p28) see page 8 and page 20

[16] "Union Official: Teachers Who Engage in Consensual Sex With Teen Pupils Shouldn't Face Prosecution" (http://www.foxnews.com/story/0,2933,432881,00.html). *Fox News*. 2008-10-06. .

[17] *Teachers International Consultancy* (2008-07-17). "Teaching at international schools is not TEFL" (http://www.ticrecruitment.com/contentpage.php?pageid=14). . Retrieved 2009-01-10.

[18] http://www.tda.gov.uk/upload/resources/pdf/t/teacher_salaries.pdf 'Teacher Salaries from September 2007' TDA (Training and Development Agency)

[19] Department of Education & Science - Education Personnel (http://www.education.ie/home/home.jsp?maincat=&pcategory=10815&ecategory=31472&sectionpage=&subject=31872&language=EN&link=&page=)

[20] Training to be a teacher (http://www.gtcs.org.uk/Becoming_a_Teacher/Trainingtobeateacher/Trainingtobeateacher.asp) GTC Scotland

[21] Teach in Scotland (http://www.teachinginscotland.com/tis/119.29.32.html)

[22] Teacher certification (http://teacherportal.com/teaching-certification)

[23] Elementary School Teachers, Except Special Education (http://www.bls.gov/oes/current/oes252021.htm)

[24] Middle School Teachers, Except Special and Vocational Education (http://www.bls.gov/oes/current/oes252022.htm)

[25] Secondary School Teachers, Except Special and Vocational Education (http://www.bls.gov/oes/current/oes252031.htm)

[26] "U.S. Department of Labor: Bureau of Labor Statistics. (July 18, 2007). *Teachers—Preschool, Kindergarten, Elementary, Middle, and Secondary: Earnings*." (http://www.bls.gov/oco/ocos069.htm#earnings). . Retrieved 2007-10-11.

[27] "U.S. Department of Labor: Bureau of Labor Statistics. (August, 2007). *Spotlight on Statistics: Back to School*." (http://www.bls.gov/spotlight/). . Retrieved 2007-10-11.

[28] "Make It Happen: A Student's Guide," (http://www.nea.org/student-program/tools/makeithappen.html) National Education Association. Retrieved 7/5/07.

[29] 2007 "Survey & Analysis of Teacher Salary Trends," (http://www.aft.org/pdfs/teachers/salarysurvey07.pdf) American Federation of Teachers. Retrieved 8/7/10.

[30] 2008 "Teacher Salary- Average Teacher Salaries" (http://www.payscale.com/research/US/All_K-12_Teachers/Salary) PayScale. Retrieved 9/16/08.

[31] "Selling Lessons Online Raises Cash and Questions" (http://www.nytimes.com/2009/11/15/education/15plans.html?_r=2&hp)

[32] "Wales' teachers 'getting younger'" (http://news.bbc.co.uk/1/hi/wales/8630117.stm). *BBC News*. 2010-04-20. .

[33] "Attacks on teachers in Wales on the rise" (http://www.bbc.co.uk/news/uk-wales-11476605). *BBC News*. 2010-10-07. .

Chemistry

Chemistry (the etymology of the word has been much disputed)[1] is the science of matter and the changes it undergoes. The science of matter is also addressed by physics, but while physics takes a more general and fundamental approach, chemistry is more specialized, being concerned with the composition, behavior (or reaction), structure, and properties of matter, as well as the changes it undergoes during chemical reactions.[2] It is a physical science which studies various substances, atoms, molecules, crystals and other aggregates of matter whether in isolation or combination, and which incorporates the concepts of energy and entropy in relation to the spontaneity of chemical processes.

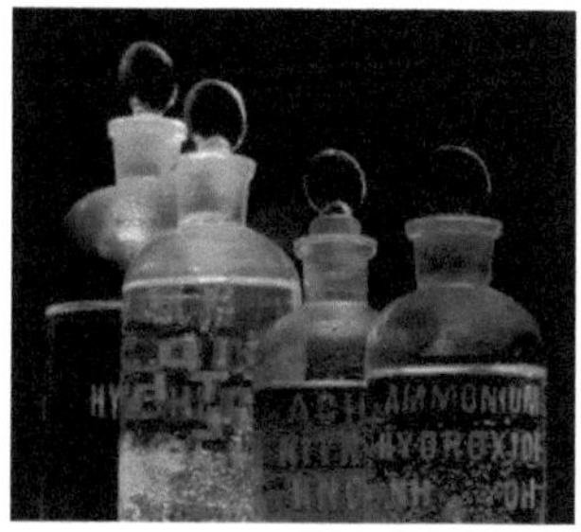

Chemistry is the science concerned with the composition, structure, and properties of matter, as well as the changes it undergoes during chemical reactions.

Disciplines within chemistry are traditionally grouped by the type of matter being studied or the kind of study. These include inorganic chemistry, the study of inorganic matter; organic chemistry, the study of organic (carbon based) matter; biochemistry, the study of substances found in biological organisms; physical chemistry, the study of chemical processes using physical concepts such as thermodynamics and quantum mechanics; and analytical chemistry, the analysis of material samples to gain an understanding of their chemical composition and structure. Many more specialized disciplines have emerged in recent years, e.g. neurochemistry the chemical study of the nervous system (see subdisciplines).

During chemical reactions, bonds between atoms break and form, resulting in different substances with different properties. In a blast furnace, iron oxide, a compound, reacts with carbon monoxide to form iron, one of the chemical elements, and carbon dioxide.

Summary

Chemistry is the scientific study of interaction of chemical substances that are constituted of atoms or the subatomic particles:[3] protons, electrons and neutrons.[4] Atoms combine to produce molecules or crystals. Chemistry is sometimes called "the central science" because it connects the other natural sciences such as astronomy, physics, material science, biology and geology.[5] [6]

The genesis of chemistry can be traced to certain practices, known as alchemy, which had been practiced for several millennia in various parts of the world, particularly the Middle East.[7]

The structure of objects we commonly use and the properties of the matter we commonly interact with are a consequence of the properties of chemical substances and their interactions. For example, steel is harder than iron because its atoms are bound together in a more rigid crystalline lattice; wood burns or undergoes rapid oxidation because it can react spontaneously with oxygen in a chemical reaction above a certain temperature; sugar and salt dissolve in water because their molecular/ionic properties are such that dissolution is preferred under the ambient conditions.

The transformations that are studied in chemistry are a result of interaction either between different chemical substances or between matter and energy. Traditional chemistry involves study of interactions between substances in a chemistry laboratory using various forms of laboratory glassware.

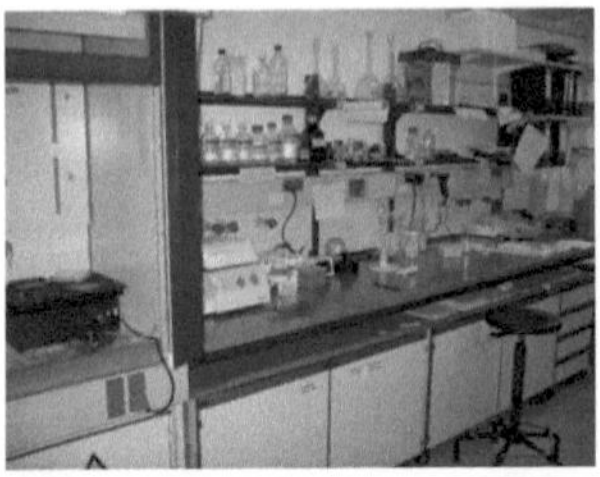

Laboratory, Institute of Biochemistry, University of Cologne

A chemical reaction is a transformation of some substances into one or more other substances.[8] It can be symbolically depicted through a chemical equation. The number of atoms on the left and the right in the equation for a chemical transformation is most often equal. The nature of chemical reactions a substance may undergo and the energy changes that may accompany it are constrained by certain basic rules, known as chemical laws.

Energy and entropy considerations are invariably important in almost all chemical studies. Chemical substances are classified in terms of their structure, phase as well as their chemical compositions. They can be analyzed using the tools of chemical analysis, e.g. spectroscopy and chromatography. Scientists engaged in chemical research are known as chemists.[9] Most chemists specialize in one or more sub-disciplines.

History

Ancient Egyptians pioneered the art of synthetic "wet" chemistry up to 4,000 years ago.[10] By 1000 BC ancient civilizations were using technologies that formed the basis of the various branches of chemistry such as; extracting metal from their ores, making pottery and glazes, fermenting beer and wine, making pigments for cosmetics and painting, extracting chemicals from plants for medicine and perfume, making cheese, dying cloth, tanning leather, rendering fat into soap, making glass, and making alloys like bronze.

The genesis of chemistry can be traced to the widely observed phenomenon of burning that led to metallurgy—the art and science of processing ores to get metals (e.g. metallurgy in ancient India). The greed for gold led to the discovery of the process for its purification, even though the underlying principles were not well understood—it was thought to be a transformation rather than purification. Many scholars in those days thought it reasonable to believe that there exist means for transforming cheaper (base) metals into gold. This gave way to alchemy and the search for the Philosopher's Stone which was believed to bring about such a transformation by mere touch.[11]

Democritus' atomist philosophy was later adopted by Epicurus (341–270 BCE).

Greek atomism dates back to 440 BC, as what might be indicated by the book *De Rerum Natura* (The Nature of Things)[12] written by the Roman Lucretius in 50 BC.[13] Much of the early development of purification methods is described by Pliny the Elder in his *Naturalis Historia.*

A tentative outline is as follows:

1. Egyptian alchemy [3,000 BCE – 400 BCE], formulate early "element" theories such as the Ogdoad.
2. Greek alchemy [332 BCE – 642 CE], the Greek king Alexander the Great conquers Egypt and founds Alexandria, having the world's largest library, where scholars and wise men gather to study.
3. Arab alchemy [642 CE – 1200], the Muslim conquest of Egypt; development of alchemy by Jābir ibn Hayyān, al-Razi and others; Jābir modifies Aristotle's theories; advances in processes and apparatus.[14]
4. European alchemy [1300 – present], Pseudo-Geber builds on Arabic chemistry. From the 12th century, major advances in the chemical arts shifted from Arab lands to western Europe.[14]
5. Chemistry [1661], Boyle writes his classic chemistry text *The Sceptical Chymist.*
6. Chemistry [1787], Lavoisier writes his classic *Elements of Chemistry.*
7. Chemistry [1803], Dalton publishes his *Atomic Theory.*
8. Chemistry [1869], Dmitry Mendeleev presented his Periodic table being the framework of the modern chemistry

The earliest pioneers of Chemistry, and inventors of the modern scientific method,[15] were medieval Arab and Persian scholars. They introduced precise observation and controlled experimentation into the field and discovered numerous Chemical substances.[16]

> "Chemistry as a science was almost created by the Muslims; for in this field, where the Greeks (so far as we know) were confined to industrial experience and vague hypothesis, the Saracens introduced precise observation, controlled experiment, and careful records. They invented and named the alembic (al-anbiq), chemically analyzed innumerable substances, composed lapidaries, distinguished alkalis and acids, investigated their affinities, studied and manufactured hundreds of drugs. Alchemy, which the Muslims inherited from Egypt, contributed to chemistry by a thousand incidental discoveries, and by its method, which was the most scientific of all medieval operations."

[16]

The most influential Muslim chemists were Jābir ibn Hayyān (Geber, d. 815), al-Kindi (d. 873), al-Razi (d. 925), al-Biruni (d. 1048) and Alhazen (d. 1039).[17] The works of Jābir became more widely known in Europe through Latin translations by a pseudo-Geber in 14th century Spain, who also wrote some of his own books under the pen name "Geber". The contribution of Indian alchemists and metallurgists in the development of chemistry was also quite significant.[18]

The emergence of chemistry in Europe was primarily due to the recurrent incidence of the plague and blights there during the so called Dark Ages. This gave rise to a need for medicines. It was thought that there exists a universal medicine called the Elixir of Life that can cure all diseases, but like the Philosopher's Stone, it was never found.

For some practitioners, alchemy was an intellectual pursuit, over time, they got better at it. Paracelsus (1493–1541), for example, rejected the 4-elemental theory and with only a vague understanding of his chemicals and medicines, formed a hybrid of alchemy and science in what was to be called *iatrochemistry*. Similarly, the influences of philosophers such as Sir Francis Bacon (1561–1626) and René Descartes (1596–1650), who demanded more rigor in mathematics and in removing bias from scientific observations, led to a scientific revolution. In chemistry, this began with Robert Boyle (1627–1691), who came up with an equation known as Boyle's Law about the characteristics of gaseous state.[20]

Antoine-Laurent de Lavoisier is considered the "Father of Modern Chemistry".[19]

Chemistry indeed came of age when Antoine Lavoisier (1743–1794), developed the theory of Conservation of mass in 1783; and the development of the Atomic Theory by John Dalton around 1800. The Law of Conservation of Mass resulted in the reformulation of chemistry based on this law and the oxygen theory of combustion, which was largely based on the work of Lavoisier. Lavoisier's fundamental contributions to chemistry were a result of a conscious effort to fit all experiments into the framework of a single theory. He established the consistent use of the chemical balance, used oxygen to overthrow the phlogiston theory, and developed a new system of chemical nomenclature and made contribution to the modern metric system. Lavoisier also worked to translate the archaic and technical language of chemistry into something that could be easily understood by the largely uneducated masses, leading to an increased public interest in chemistry. All these advances in chemistry led to what is usually called the chemical revolution. The contributions of Lavoisier led to what is now called modern chemistry—the chemistry that is studied in educational institutions all over the world. It is because of these and other contributions that Antoine Lavoisier is often celebrated as the "Father of Modern Chemistry".[21] The later discovery of Friedrich Wöhler that many natural substances, organic compounds, can indeed be synthesized in a chemistry laboratory also helped the modern chemistry to mature from its infancy.[22]

The discovery of the chemical elements has a long history from the days of alchemy and culminating in the discovery of the periodic table of the chemical elements by Dmitri Mendeleev (1834–1907)[23] and later discoveries of some synthetic elements.

Etymology

The word *chemistry* comes from the word *alchemy*, an earlier set of practices that encompassed elements of chemistry, metallurgy, philosophy, astrology, astronomy, mysticism and medicine; it is commonly thought of as the quest to turn lead or another common starting material into gold.[24] The word *alchemy* in turn is derived from the Arabic word al-kīmīā (الكيمياء), meaning alchemy. The Arabic term is borrowed from the Greek χημία or χημεία.[25] [26] This may have Egyptian origins. Many believe that al-kīmīā is derived from χημία, which is in turn derived from the word **Chemi** or **Kimi**, which is the ancient name of Egypt in Egyptian.[25] Alternately, al-kīmīā may be derived from χημεία, meaning "cast together".[27]

An alchemist was called a 'chemist' in popular speech, and later the suffix "-ry" was added to this to describe the art of the chemist as "chemistry".

Definitions

In retrospect, the definition of chemistry has changed over time, as new discoveries and theories add to the functionality of the science. Shown below are some of the standard definitions used by various noted chemists:

- **Alchemy** (330) – the study of the composition of waters, movement, growth, embodying, disembodying, drawing the spirits from bodies and bonding the spirits within bodies (Zosimos).[28]
- **Chymistry** (1661) – the subject of the material principles of mixed bodies (Boyle).[29]
- **Chymistry** (1663) – a scientific art, by which one learns to dissolve bodies, and draw from them the different substances on their composition, and how to unite them again, and exalt them to a higher perfection (Glaser).[30]
- **Chemistry** (1730) – the art of resolving mixt, compound, or aggregate bodies into their principles; and of composing such bodies from those principles (Stahl).[31]
- **Chemistry** (1837) – the science concerned with the laws and effects of molecular forces (Dumas).[32]
- **Chemistry** (1947) – the science of substances: their structure, their properties, and the reactions that change them into other substances (Pauling).[33]
- **Chemistry** (1998) – the study of matter and the changes it undergoes (Chang).[34]

Basic concepts

Several concepts are essential for the study of chemistry; some of them are:[35]

Atom

An *atom* is the basic unit of chemistry. It consists of a positively charged core (the atomic nucleus) which contains protons and neutrons, and which maintains a number of electrons to balance the positive charge in the nucleus. The atom is also the smallest entity that can be envisaged to retain some of the chemical properties of the element, such as electronegativity, ionization potential, preferred oxidation state(s), coordination number, and preferred types of bonds to form (e.g., metallic, ionic, covalent).

Element

The concept of *chemical element* is related to that of chemical substance. A chemical element is specifically a substance which is composed of a single type of atom. A chemical element is characterized by a particular number of protons in the nuclei of its atoms. This number is known as the atomic number of the element. For example, all atoms with 6 protons in their nuclei are atoms of the chemical element carbon, and all atoms with 92 protons in their nuclei are atoms of the element uranium. Ninety–four different chemical elements or types of atoms based on the number of protons exist naturally. A further 18 have been recognised by IUPAC as existing artificially only. Although all the nuclei of all atoms belonging to one element will have the same number of protons, they may not necessarily have the same number of neutrons; such atoms are termed isotopes. In fact several isotopes of an element may exist.

The most convenient presentation of the chemical elements is in the periodic table of the chemical elements, which groups elements by atomic number. Due to its ingenious arrangement, groups, or columns, and periods, or rows, of elements in the table either share several chemical properties, or follow a certain trend in characteristics such as atomic radius, electronegativity, etc. Lists of the elements by name, by symbol, and by atomic number are also available.

Compound

A *compound* is a substance with a *particular ratio* of atoms of particular chemical elements which determines its composition, and a particular organization which determines chemical properties. For example, water is a compound containing hydrogen and oxygen in the ratio of two to one, with the oxygen atom between the two hydrogen atoms, and an angle of 104.5° between them. Compounds are formed and interconverted by chemical reactions.

Substance

A chemical substance is a kind of matter with a definite composition and set of properties.[36] Strictly speaking, a mixture of compounds, elements or compounds and elements is not a chemical substance, but it may be called a chemical. Most of the substances we encounter in our daily life are some kind of mixture; for example: air, alloys, biomass, etc.

Nomenclature of substances is a critical part of the language of chemistry. Generally it refers to a system for naming chemical compounds. Earlier in the history of chemistry substances were given name by their discoverer, which often led to some confusion and difficulty. However, today the IUPAC system of chemical nomenclature allows chemists to specify by name specific compounds amongst the vast variety of possible chemicals. The standard nomenclature of chemical substances is set by the International Union of Pure and Applied Chemistry (IUPAC). There are well-defined systems in place for naming chemical species. Organic compounds are named according to the organic nomenclature system.[37] Inorganic compounds are named according to the inorganic nomenclature system.[38] In addition the Chemical Abstracts Service has devised a method to index chemical substance. In this scheme each chemical substance is identifiable by a number known as CAS registry number.

Molecule

A *molecule* is the smallest indivisible portion of a pure chemical substance that has its unique set of chemical properties, that is, its potential to undergo a certain set of chemical reactions with other substances. Molecules can exist as electrically neutral units unlike ions. Molecules are typically a set of atoms bound together by covalent bonds, such that the structure is electrically neutral and all valence electrons are paired with other electrons either in bonds or in lone pairs.

Not all substances consist of discrete molecules. Most chemical elements are composed of lone atoms as their smallest discrete unit. Other types of substances, such as ionic compounds and network solids, are organized in such a way as to lack the existence of identifiable molecules *per se*. Instead, these substances are discussed in terms of formula units or unit cells as the smallest repeating structure within the substance; as they lack identifiable molecules.

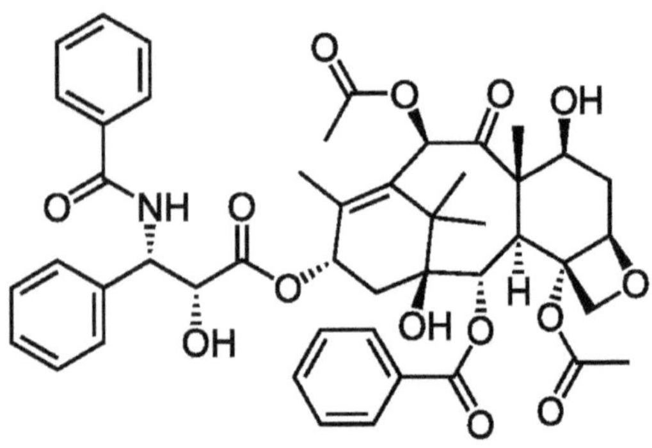

A molecular structure depicts the bonds and relative positions of atoms in a molecule such as that in Paclitaxel shown here

One of the main characteristic of a molecule is its geometry often called its structure. While the structure of diatomic, triatomic or tetra atomic molecules may be trivial, (linear, angular pyramidal etc.) the structure of polyatomic molecules, that are constituted of more than six atoms (of several elements) can be crucial for its chemical nature.

Mole and amount of substance

Mole is a unit to measure amount of substance (also called chemical amount). A mole is the amount of a substance that contains as many elementary entities (atoms, molecules or ions) as there are atoms in 0.012 kilogram (or 12 grams) of carbon-12, where the carbon-12 atoms are unbound, at rest and in their ground state.[39] The number of entities per mole is known as the Avogadro constant, and is determined empirically. The currently accepted value is $6.02214179(30)\times10^{23}$ mol^{-1} (2007 CODATA). One way to understand the meaning of the term "mole" is to compare and contrast it to terms such as dozen. Just as one dozen eggs contains 12 individual eggs, one mole contains $6.02214179(30)\times10^{23}$ atoms, molecules or other particles. The term is used because it is much easier to say, for example, 1 mole of carbon, than it is to say $6.02214179(30)\times10^{23}$ carbon atoms, and because moles of chemicals represent a scale that is easy to experience.

The amount of substance of a solute per volume of solution is known as amount of substance concentration, or molarity for short. Molarity is the quantity most commonly used to express the concentration of a solution in the chemical laboratory. The most commonly used units for molarity are mol/L (the official SI units are mol/m^3).

Ions and salts

An *ion* is a charged species, an atom or a molecule, that has lost or gained one or more electrons. Positively charged cations (e.g. sodium cation Na^+) and negatively charged anions (e.g. chloride Cl^-) can form a crystalline lattice of neutral salts (e.g. sodium chloride NaCl). Examples of polyatomic ions that do not split up during acid-base reactions are hydroxide (OH^-) and phosphate (PO_4^{3-}).

Ions in the gaseous phase are often known as plasma.

Acidity and basicity

A substance can often be classified as an acid or a base. There are several different theories which explain acid-base behavior. The simplest is Arrhenius theory, which states than an acid is a substance that produces hydronium ions when it is dissolved in water, and a base is one that produces hydroxide ions when dissolved in water. According to Brønsted–Lowry acid-base theory, acids are substances that donate a positive hydrogen ion to another substance in a chemical reaction; by extension, a base is the substance which receives that hydrogen ion. A third common theory is Lewis acid-base theory, which is based on the formation of new chemical bonds. Lewis theory explains that an acid is a substance which is capable of accepting a pair of electrons from another substance during the process of bond formation, while a base is a substance which can provide a pair of electrons to form a new bond. According to concept as per Lewis, the crucial things being exchanged are charges.[40] There are several other ways in which a substance may be classified as an acid or a base, as is evident in the history of this concept [41]

Acid strength is commonly measured by two methods. One measurement, based on the Arrhenius definition of acidity, is pH, which is a measurement of the hydronium ion concentration in a solution, as expressed on a negative logarithmic scale. Thus, solutions that have a low pH have a high hydronium ion concentration, and can be said to be more acidic. The other measurement, based on the Brønsted–Lowry definition, is the acid dissociation constant (K_a), which measure the relative ability of a substance to act as an acid under the Brønsted–Lowry definition of an acid. That is, substances with a higher K_a are more likely to donate hydrogen ions in chemical reactions than those with lower K_a values.

Phase

In addition to the specific chemical properties that distinguish different chemical classifications chemicals can exist in several phases. For the most part, the chemical classifications are independent of these bulk phase classifications; however, some more exotic phases are incompatible with certain chemical properties. A *phase* is a set of states of a chemical system that have similar bulk structural properties, over a range of conditions, such as pressure or temperature. Physical properties, such as density and refractive index tend to fall within values characteristic of the phase. The phase of matter is defined by the *phase transition*, which is when energy put into or taken out of the system goes into rearranging the structure of the system, instead of changing the bulk conditions.

Sometimes the distinction between phases can be continuous instead of having a discrete boundary, in this case the matter is considered to be in a supercritical state. When three states meet based on the conditions, it is known as a triple point and since this is invariant, it is a convenient way to define a set of conditions.

The most familiar examples of phases are solids, liquids, and gases. Many substances exhibit multiple solid phases. For example, there are three phases of solid iron (alpha, gamma, and delta) that vary based on temperature and pressure. A principal difference between solid phases is the crystal structure, or arrangement, of the atoms. Another phase commonly encountered in the study of chemistry is the *aqueous* phase, which is the state of substances dissolved in aqueous solution (that is, in water). Less familiar phases include plasmas, Bose-Einstein condensates and fermionic condensates and the paramagnetic and ferromagnetic phases of magnetic materials. While most familiar phases deal with three-dimensional systems, it is also possible to define analogs in two-dimensional systems, which has received attention for its relevance to systems in biology.

Redox

It is a concept related to the ability of atoms of various substances to lose or gain electrons. Substances that have the ability to oxidize other substances are said to be oxidative and are known as oxidizing agents, oxidants or oxidizers. An oxidant removes electrons from another substance. Similarly, substances that have the ability to reduce other substances are said to be reductive and are known as reducing agents, reductants, or reducers. A reductant transfers electrons to another substance, and is thus oxidized itself. And because it "donates" electrons it is also called an electron donor. Oxidation and reduction properly refer to a change in oxidation number—the actual transfer of electrons may never occur. Thus, oxidation is better defined as an increase in oxidation number, and reduction as a decrease in oxidation number.

Bonding

Atoms sticking together in molecules or crystals are said to be bonded with one another. A chemical bond may be visualized as the multipole balance between the positive charges in the nuclei and the negative charges oscillating about them.[42] More than simple attraction and repulsion, the energies and distributions characterize the availability of an electron to bond to another atom.

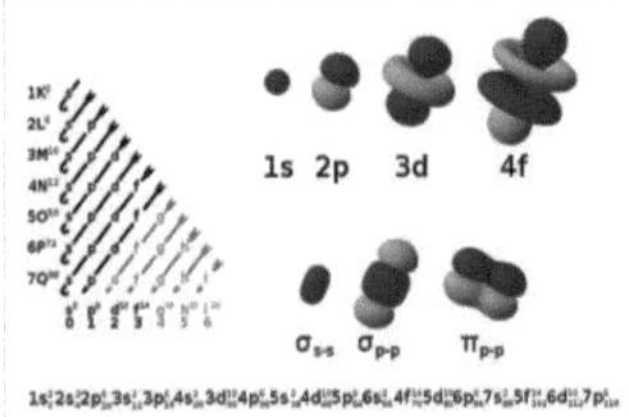

Electron atomic and molecular orbitals

A chemical bond can be a covalent bond, an ionic bond, a hydrogen bond or just because of Van der Waals force. Each of these kind of bond is ascribed to some potential. These potentials create the interactions which hold atoms together in molecules or crystals. In many simple compounds, Valence Bond Theory, the Valence Shell Electron Pair Repulsion model (VSEPR), and the concept of oxidation number can be used to explain molecular structure and composition. Similarly, theories from classical physics can be used to predict many ionic structures. With more complicated compounds, such as metal complexes, valence bond theory is less applicable and alternative approaches, such as the molecular orbital theory, are generally used. See diagram on electronic orbitals.

Reaction

When a chemical substance is transformed as a result of its interaction with another or energy, a chemical reaction is said to have occurred. *Chemical reaction* is therefore a concept related to the 'reaction' of a substance when it comes in close contact with another, whether as a mixture or a solution; exposure to some form of energy, or both. It results in some energy exchange between the constituents of the reaction as well with the system environment which may be a designed vessels which are often laboratory glassware. Chemical reactions can result in the formation or dissociation of molecules, that is, molecules breaking apart to form two or more smaller molecules, or rearrangement of atoms within or across molecules. Chemical reactions usually involve the making or breaking of chemical bonds. Oxidation, reduction, dissociation, acid-base neutralization and molecular rearrangement are some of the commonly used kinds of chemical reactions.

A chemical reaction can be symbolically depicted through a chemical equation. While in a non-nuclear chemical reaction the number and kind of atoms on both sides of the equation are equal, for a nuclear reaction this holds true only for the nuclear particles viz. protons and neutrons.[43]

The sequence of steps in which the reorganization of chemical bonds may be taking place in the course of a chemical reaction is called its mechanism. A chemical reaction can be envisioned to take place in a number of steps, each of which may have a different speed. Many reaction intermediates with variable stability can thus be envisaged during the course of a reaction. Reaction mechanisms are proposed to explain the kinetics and the relative product mix of a reaction. Many physical chemists specialize in exploring and proposing the mechanisms of various chemical reactions. Several empirical rules, like the Woodward-Hoffmann rules often come handy while proposing a mechanism for a chemical reaction.

According to the IUPAC gold book a chemical reaction is a process that results in the interconversion of chemical species".[44] Accordingly, a chemical reaction may be an elementary reaction or a stepwise reaction. An additional caveat is made, in that this definition includes cases where the interconversion of conformers is experimentally observable. Such detectable chemical reactions normally involve sets of molecular entities as indicated by this definition, but it is often conceptually convenient to use the term also for changes involving single molecular entities (i.e. 'microscopic chemical events').

Equilibrium

Although the concept of equilibrium is widely used across sciences, in the context of chemistry, it arises whenever a number of different states of the chemical composition are possible. For example, in a mixture of several chemical compounds that can react with one another, or when a substance can be present in more than one kind of phase. A system of chemical substances at equilibrium even though having an unchanging composition is most often not static; molecules of the substances continue to react with one another thus giving rise to a dynamic equilibrium. Thus the concept describes the state in which the parameters such as chemical composition remain unchanged over time. Chemicals present in biological systems are invariably not at equilibrium; rather they are far from equilibrium.

Energy

In the context of chemistry, energy is an attribute of a substance as a consequence of its atomic, molecular or aggregate structure. Since a chemical transformation is accompanied by a change in one or more of these kinds of structure, it is invariably accompanied by an increase or decrease of energy of the substances involved. Some energy is transferred between the surroundings and the reactants of the reaction in the form of heat or light; thus the products of a reaction may have more or less energy than the reactants. A reaction is said to be exergonic if the final state is lower on the energy scale than the initial state; in the case of endergonic reactions the situation is the reverse. A reaction is said to be exothermic if the reaction releases heat to the surroundings; in the case of endothermic reactions, the reaction absorbs heat from the surroundings.

Chemical reactions are invariably not possible unless the reactants surmount an energy barrier known as the activation energy. The *speed* of a chemical reaction (at given temperature T) is related to the activation energy E, by the Boltzmann's population factor - that is the probability of molecule to have energy greater than or equal to E at the given temperature T. This exponential dependence of a reaction rate on temperature is known as the Arrhenius equation. The activation energy necessary for a chemical reaction can be in the form of heat, light, electricity or mechanical force in the form of ultrasound.[45]

A related concept free energy, which also incorporates entropy considerations, is a very useful means for predicting the feasibility of a reaction and determining the state of equilibrium of a chemical reaction, in chemical thermodynamics. A reaction is feasible only if the total change in the Gibbs free energy is negative, ; if it is equal to zero the chemical reaction is said to be at equilibrium.

There exist only limited possible states of energy for electrons, atoms and molecules. These are determined by the rules of quantum mechanics, which require quantization of energy of a bound system. The atoms/molecules in a higher energy state are said to be excited. The molecules/atoms of substance in an excited energy state are often much more reactive; that is, more amenable to chemical reactions.

The phase of a substance is invariably determined by its energy and the energy of its surroundings. When the intermolecular forces of a substance are such that the energy of the surroundings is not sufficient to overcome them, it occurs in a more ordered phase like liquid or solid as is the case with water (H_2O); a liquid at room temperature because its molecules are bound by hydrogen bonds.[46] Whereas hydrogen sulfide (H_2S) is a gas at room temperature and standard pressure, as its molecules are bound by weaker dipole-dipole interactions.

The transfer of energy from one chemical substance to another depends on the *size* of energy quanta emitted from one substance. However, heat energy is often transferred more easily from almost any substance to another because the phonons responsible for vibrational and rotational energy levels in a substance have much less energy than photons invoked for the electronic energy transfer. Thus, because vibrational and rotational energy levels are more closely spaced than electronic energy levels, heat is more easily transferred between substances relative to light or other forms of electronic energy. For example, ultraviolet electromagnetic radiation is not transferred with as much efficacy from one substance to another as thermal or electrical energy.

The existence of characteristic energy levels for different chemical substances is useful for their identification by the analysis of spectral lines. Different kinds of spectra are often used in chemical spectroscopy, e.g. IR, microwave, NMR, ESR, etc. Spectroscopy is also used to identify the composition of remote objects - like stars and distant galaxies - by analyzing their radiation spectra.

Emission spectrum of iron

The term chemical energy is often used to indicate the potential of a chemical substance to undergo a transformation through a chemical reaction or to transform other chemical substances.

Chemical laws

Chemical reactions are governed by certain laws, which have become fundamental concepts in chemistry. Some of them are:

- Avogadro's law
- Beer-Lambert law
- Boyle's law (1662, relating pressure and volume)
- Charles's law (1787, relating volume and temperature)
- Fick's law of diffusion
- Gay-Lussac's law (1809, relating pressure and temperature)
- Le Chatelier's Principle
- Henry's law
- Hess's Law
- Law of conservation of energy leads to the important concepts of equilibrium, thermodynamics, and kinetics.
- Law of conservation of mass, according to the modern physics it is actually energy that is conserved, and that energy and mass are related; a concept which becomes important in nuclear chemistry.
- Law of definite composition, although in many systems (notably biomacromolecules and minerals) the ratios tend to require large numbers, and are frequently represented as a fraction.
- Law of multiple proportions
- Raoult's Law

Subdisciplines

Chemistry is typically divided into several major sub-disciplines. There are also several main cross-disciplinary and more specialized fields of chemistry.[47]

- Analytical chemistry is the analysis of material samples to gain an understanding of their chemical composition and structure. Analytical chemistry incorporates standardized experimental methods in chemistry. These methods may be used in all subdisciplines of chemistry, excluding purely theoretical chemistry.
- Biochemistry is the study of the chemicals, chemical reactions and chemical interactions that take place in living organisms. Biochemistry and organic chemistry are closely related, as in medicinal chemistry or neurochemistry. Biochemistry is also associated with molecular biology and genetics.
- Inorganic chemistry is the study of the properties and reactions of inorganic compounds. The distinction between organic and inorganic disciplines is not absolute and there is much overlap, most importantly in the sub-discipline of organometallic chemistry.
- Materials chemistry is the preparation, characterization, and understanding of substances with a useful function. The field is a new breadth of study in graduate programs, and it integrates elements from all classical areas of

chemistry with a focus on fundamental issues that are unique to materials. Primary systems of study include the chemistry of condensed phases (solids, liquids, polymers) and interfaces between different phases.

- Neurochemistry is the study of neurochemicals; including transmitters, peptides, proteins, lipids, sugars, and nucleic acids; their interactions, and the roles they play in forming, maintaining, and modifying the nervous system.
- Nuclear chemistry is the study of how subatomic particles come together and make nuclei. Modern Transmutation is a large component of nuclear chemistry, and the table of nuclides is an important result and tool for this field.
- Organic chemistry is the study of the structure, properties, composition, mechanisms, and reactions of organic compounds. An organic compound is defined as any compound based on a carbon skeleton.
- Physical chemistry is the study of the physical and fundamental basis of chemical systems and processes. In particular, the energetics and dynamics of such systems and processes are of interest to physical chemists. Important areas of study include chemical thermodynamics, chemical kinetics, electrochemistry, statistical mechanics, spectroscopy, and more recently, astrochemistry.[48] Physical chemistry has large overlap with molecular physics. Physical chemistry involves the use of infinitesimal calculus in deriving equations. It is usually associated with quantum chemistry and theoretical chemistry. Physical chemistry is a distinct discipline from chemical physics, but again, there is very strong overlap.
- Theoretical chemistry is the study of chemistry via fundamental theoretical reasoning (usually within mathematics or physics). In particular the application of quantum mechanics to chemistry is called quantum chemistry. Since the end of the Second World War, the development of computers has allowed a systematic development of computational chemistry, which is the art of developing and applying computer programs for solving chemical problems. Theoretical chemistry has large overlap with (theoretical and experimental) condensed matter physics and molecular physics.

Other fields include agrochemistry, astrochemistry (and cosmochemistry), atmospheric chemistry, chemical engineering, chemical biology, chemo-informatics, electrochemistry, environmental chemistry, femtochemistry, flavor chemistry, flow chemistry, geochemistry, green chemistry, histochemistry, history of chemistry, hydrogenation chemistry, immunochemistry, marine chemistry, materials science, mathematical chemistry, mechanochemistry, medicinal chemistry, molecular biology, molecular mechanics, nanotechnology, natural product chemistry, oenology, organometallic chemistry, petrochemistry, pharmacology, photochemistry, physical organic chemistry, phytochemistry, polymer chemistry, radiochemistry, solid-state chemistry, sonochemistry, supramolecular chemistry, surface chemistry, synthetic chemistry, thermochemistry, and many others.

Chemical industry

The chemical industry represents an important economic activity. The global top 50 chemical producers in 2004 had sales of 587 billion US dollars with a profit margin of 8.1% and research and development spending of 2.1% of total chemical sales.[49]

Professional societies

- American Chemical Society
- American Society for Neurochemistry
- Chemical Institute of Canada
- Chemical Society of Peru
- International Union of Pure and Applied Chemistry
- Royal Australian Chemical Institute
- Royal Netherlands Chemical Society
- Royal Society of Chemistry

- Society of Chemical Industry
- World Association of Theoretical and Computational Chemists
- List of chemistry societies

See also

- Common chemicals
- International Year of Chemistry
- List of chemists
- List of compounds
- List of important publications in chemistry
- List of software for molecular mechanics modeling
- List of unsolved problems in chemistry
- Philosophy of chemistry

References

[1] **See:** Chemistry (etymology) for possible origins of this word.
[2] Chemistry (http://dictionary.reference.com/browse/Chemistry). (n.d.). Merriam-Webster's Medical Dictionary. Retrieved August 19, 2007.
[3] "What is Chemistry?" (http://chemweb.ucc.ie/what_is_chemistry.htm). Chemweb.ucc.ie. . Retrieved 2011-06-12.
[4] Matter: Atoms from Democritus to Dalton (http://www.visionlearning.com/library/module_viewer.php?mid=49) by Anthony Carpi, Ph.D.
[5] Theodore L. Brown, H. Eugene Lemay, Bruce Edward Bursten, H. Lemay. *Chemistry: The Central Science*. Prentice Hall; 8 edition (1999). ISBN 0-13-010310-1. Pages 3-4.
[6] It is sometimes called the central science because it is seen as occupying an intermediate position in a hierarchy of the sciences by "reductive level", between physics and biology. See Carsten Reinhardt. *Chemical Sciences in the 20th Century: Bridging Boundaries*. Wiley-VCH, 2001. ISBN 3-527-30271-9. Pages 1-2.
[7] http://etext.lib.virginia.edu/cgi-local/DHI/dhi.cgi?id=dv1-04
[8] IUPAC Gold Book Definition (http://www.iupac.org/goldbook/C01033.pdf)
[9] "California Occupational Guide Number 22: Chemists" (http://www.calmis.ca.gov/file/occguide/CHEMIST.HTM). Calmis.ca.gov. 1999-10-29. . Retrieved 2011-06-12.
[10] First chemists (http://www.newscientist.com/article/mg16121734.300-first-chemists.html), February 13, 1999, New Scientist
[11] Alchemy Timeline (http://www.chemheritage.org/explore/ancients-time.html) - Chemical Heritage Society
[12] Lucretius (50 BCE). "de Rerum Natura (On the Nature of Things)" (http://classics.mit.edu/Carus/nature_things.html). *The Internet Classics Archive*. Massachusetts Institute of Technology. . Retrieved 2007-01-09.
[13] Simpson, David (29 June 2005). "Lucretius (c. 99 - c. 55 BCE)" (http://www.iep.utm.edu/l/lucretiu.htm). *The Internet History of Philosophy*. . Retrieved 2007-01-09.
[14] Richard Myers (2003). " *The Basics of Chemistry* (http://books.google.com/books?id=oS50J3-IfZsC&pg=PA13&dq&hl=en#v=onepage&q=&f=false)". Greenwood Publishing Group. pp.13–14. ISBN 0-313-31664-3
[15] Morris Kline (1985) *Mathematics for the nonmathematician* (http://books.google.com/books?id=f-e0bro-0FUC&pg=PA284&dq&hl=en#v=onepage&q=&f=false). Courier Dover Publications. p. 284. ISBN 0-486-24823-2
[16] Will Durant (1980), *The Age of Faith (The Story of Civilization, Volume 4)*, p. 162-186, Simon & Schuster, ISBN 0-671-01200-2
[17] Dr. K. Ajram (1992), *Miracle of Islamic Science*, Appendix B, Knowledge House Publishers, ISBN 0-911119-43-4.

"Humboldt regards the Muslims as the founders of chemistry."

[18] Will Durant (1935): Our Oriental Heritage: Simon & Schuster:

"Something has been said about the chemical excellence of cast iron in ancient India, and about the high industrial development of the Gupta times, when India was looked to, even by Imperial Rome, as the most skilled of the nations in such chemical industries as dyeing, tanning, soap-making, glass and cement... By the sixth century the Hindus were far ahead of Europe in industrial chemistry; they were masters of calcination, distillation, sublimation, steaming, fixation, the production of light without heat, the mixing of anesthetic and soporific powders, and the preparation of metallic salts, compounds and alloys. The tempering of steel was brought in ancient India to a perfection unknown in Europe till our own times; King Porus is said to have selected, as a specially valuable gift from Alexander, not gold or silver, but thirty pounds of steel. The Moslems took much of this Hindu chemical science and industry to the Near East and Europe; the secret of manufacturing "Damascus" blades, for example, was taken by the Arabs from the Persians, and by the Persians from India.""

[19] Eagle, Cassandra T.; Jennifer Sloan (1998). "Marie Anne Paulze Lavoisier: The Mother of Modern Chemistry" (http://www.springerlink.com/content/x14v35m5n8822v42/fulltext.pdf) (PDF). *The Chemical Educator* **3** (5): 1 – 18. doi:10.1007/s00897980249a. . Retrieved 2007-12-14.

[20] "History - Robert Boyle (1627 - 1691)" (http://www.bbc.co.uk/history/historic_figures/boyle_robert.shtml). BBC. . Retrieved 2011-06-12.

[21] Mi Gyung Kim (2003). *Affinity, that Elusive Dream: A Genealogy of the Chemical Revolution*. MIT Press. p. 440. ISBN 0262112736.

[22] Ihde, Aaron John (1984). *The Development of Modern Chemistry*. Courier Dover Publications. p. 164. ISBN 0486642356.

[23] Timeline of Element Discovery (http://chemistry.about.com/library/weekly/aa030303a.htm) - About.com

[24] "History of Alchemy" (http://www.alchemylab.com/history_of_alchemy.htm). Alchemy Lab. . Retrieved 2011-06-12.

[25] "alchemy", entry in *The Oxford English Dictionary*, J. A. Simpson and E. S. C. Weiner, vol. 1, 2nd ed., 1989, ISBN 0-19-861213-3.

[26] p. 854, "Arabic alchemy", Georges C. Anawati, pp. 853-885 in *Encyclopedia of the history of Arabic science*, eds. Roshdi Rashed and Régis Morelon, London: Routledge, 1996, vol. 3, ISBN 0415124123.

[27] Weekley, Ernest (1967). Etymological Dictionary of Modern English. New York: Dover Publications. ISBN 0486218732

[28] Strathern, P. (2000). *Mendeleyev's Dream – the Quest for the Elements*. New York: Berkley Books.

[29] Boyle, Robert (1661). *The Sceptical Chymist*. New York: Dover Publications, Inc. (reprint). ISBN 0486428257.

[30] Glaser, Christopher (1663). *Traite de la chymie*. Paris. as found in: Kim, Mi Gyung (2003). *Affinity, That Elusive Dream - A Genealogy of the Chemical Revolution*. The MIT Press. ISBN 0-262-11273-6.

[31] Stahl, George, E. (1730). *Philosophical Principles of Universal Chemistry*. London.

[32] Dumas, J. B. (1837). 'Affinite' (lecture notes), vii, pg 4. "Statique chimique", Paris: Academie des Sciences

[33] Pauling, Linus (1947). *General Chemistry*. Dover Publications, Inc.. ISBN 0486656225.

[34] Chang, Raymond (1998). *Chemistry, 6th Ed.*. New York: McGraw Hill. ISBN 0-07-115221-0.

[35] "General Chemistry Online - Companion Notes: Matter" (http://antoine.frostburg.edu/chem/senese/101/matter/). Antoine.frostburg.edu. . Retrieved 2011-06-12.

[36] Hill, J.W.; Petrucci, R.H.; McCreary, T.W.; Perry, S.S. (2005). *General Chemistry* (4th ed.). Upper Saddle River, NJ: Pearson Prentice Hall. p. 37.

[37] "IUPAC Nomenclature of Organic Chemistry" (http://www.acdlabs.com/iupac/nomenclature/). Acdlabs.com. . Retrieved 2011-06-12.

[38] IUPAC Provisional Recommendations for the Nomenclature of Inorganic Chemistry (2004) (http://www.iupac.org/reports/provisional/abstract04/connelly_310804.html)

[39] "Official SI Unit definitions" (http://www.bipm.org/en/si/base_units/). Bipm.org. . Retrieved 2011-06-12.

[40] "The Lewis Acid-Base Concept" (http://web.archive.org/web/20080527132328/http://www.apsidium.com/theory/lewis_acid.htm). *Apsidium*. May 19, 2003. Archived from the original (http://www.apsidium.com/theory/lewis_acid.htm) on 2008-05-27. . Retrieved 2010-07-31.

[41] "History of Acidity" (http://www.bbc.co.uk/dna/h2g2/A708257). Bbc.co.uk. 2004-05-27. . Retrieved 2011-06-12.

[42] Visionlearning. "Chemical Bonding by Anthony Carpi, Ph" (http://www.visionlearning.com/library/module_viewer.php?mid=55). visionlearning. . Retrieved 2011-06-12.

[43] Chemical Reaction Equation (http://goldbook.iupac.org/C01034.html)- IUPAC Goldbook

[44] Gold Book Chemical Reaction (http://goldbook.iupac.org/C01033.html) IUPAC Goldbook

[45] Reilly, Michael. (2007). Mechanical force induces chemical reaction (http://www.newscientisttech.com/article/dn11427), NewScientist.com news service, Reilly

[46] Changing States of Matter (http://www.chem4kids.com/files/matter_changes.html) - Chemforkids.com

[47] W.G. Laidlaw; D.E. Ryan And Gary Horlick; H.C. Clark, Josef Takats, And Martin Cowie; R.U. Lemieux (1986-12-10). "Chemistry Subdisciplines" (http://www.thecanadianencyclopedia.com/index.cfm?PgNm=TCE&Params=A1ARTA0001555). The Canadian Encyclopedia. . Retrieved 2011-06-12.

[48] Herbst, Eric (May 12, 2005). "Chemistry of Star-Forming Regions". *Journal of Physical Chemistry A* **109** (18): 4017–4029. doi:10.1021/jp050461c. PMID 16833724.

[49] "Top 50 Chemical Producers" (http://pubs.acs.org/cen/coverstory/83/8329globaltop50.html). *Chemical & Engineering News* **83** (29): 20–23. July 18, 2005. .

Further reading

Popular reading

- Atkins, P.W. *Galileo's Finger* (Oxford University Press) ISBN 0-19-860941-8
- Atkins, P.W. *Atkins' Molecules* (Cambridge University Press) ISBN 0-521-82397-8
- Stwertka, A. *A Guide to the Elements* (Oxford University Press) ISBN 0-19-515027-9

Introductory undergraduate text books

- Atkins, P.W., Overton, T., Rourke, J., Weller, M. and Armstrong, F. *Shriver and Atkins inorganic chemistry* (4th edition) 2006 (Oxford University Press) ISBN 0-19-926463-5
- Chang, Raymond. *Chemistry* 6th ed. Boston: James M. Smith, 1998. ISBN 0-07-115221-0.
- Clayden, Jonathan; Greeves, Nick; Warren, Stuart; Wothers, Peter (2001). *Organic Chemistry* (1st ed.). Oxford University Press. ISBN 978-0-19-850346-0.
- Voet and Voet *Biochemistry* (Wiley) ISBN 0-471-58651-X

Advanced undergraduate-level or graduate text books

- Atkins, P.W. *Physical Chemistry* (Oxford University Press) ISBN 0-19-879285-9
- Atkins, P.W. et al. *Molecular Quantum Mechanics* (Oxford University Press)
- McWeeny, R. *Coulson's Valence* (Oxford Science Publications) ISBN 0-19-855144-4
- Pauling, L. *The Nature of the chemical bond* (Cornell University Press) ISBN 0-8014-0333-2
- Pauling, L., and Wilson, E. B. *Introduction to Quantum Mechanics with Applications to Chemistry* (Dover Publications) ISBN 0-486-64871-0
- Smart and Moore *Solid State Chemistry: An Introduction* (Chapman and Hall) ISBN 0-412-40040-5
- Stephenson, G. *Mathematical Methods for Science Students* (Longman) ISBN 0-582-44416-0

bjn:Kimia rue:Хемія

Analytical chemistry

Analytical chemistry is the study of the separation, identification, and quantification of the chemical components of natural and artificial materials.[1] Qualitative analysis gives an indication of the identity of the chemical species in the sample and quantitative analysis determines the amount of one or more of these components. The separation of components is often performed prior to analysis.

Analytical methods can be separated into classical and instrumental.[2] Classical methods (also known as wet chemistry methods) use separations such as precipitation, extraction, and distillation and qualitative analysis by color, odor, or melting point. Quantitative analysis is achieved by measurement of weight or volume. Instrumental methods use an apparatus to measure physical quantities of the analyte such as light absorption, fluorescence, or conductivity. The separation of materials is accomplished using chromatography or electrophoresis methods.

Analytical chemistry is also focused on improvements in experimental design, chemometrics, and the creation of new measurement tools to provide better chemical information. Analytical chemistry has applications in forensics, bioanalysis, clinical analysis, environmental analysis, and materials analysis.

History

Analytical chemistry has been important since the early days of chemistry, providing methods for determining which elements and chemicals are present in the world around us. During this period significant analytical contributions to chemistry include the development of systematic elemental analysis by Justus von Liebig and systematized organic analysis based on the specific reactions of functional groups.

Gustav Kirchhoff (left) and Robert Bunsen (right)

The first instrumental analysis was flame emissive spectrometry developed by Robert Bunsen and Gustav Kirchhoff who discovered rubidium (Rb) and caesium (Cs) in 1860.[3]

Most of the major developments in analytical chemistry take place after 1900. During this period instrumental analysis becomes progressively dominant in the field. In particular many of the basic spectroscopic and spectrometric techniques were discovered in the early 20th century and refined in the late 20th century.[4]

The separation sciences follow a similar time line of development and also become increasingly transformed into high performance instruments.[5] In the 1970s many of these techniques began to be used together to achieve a complete characterization of samples.

Starting in approximately the 1970s into the present day analytical chemistry has progressively become more inclusive of biological questions (bioanalytical chemistry), whereas it had previously been largely focused on inorganic or small organic molecules. Lasers have been increasingly used in chemistry as probes and even to start and influence a wide variety of reactions. The late 20th century also saw an expansion of the application of analytical chemistry from somewhat academic chemical questions to forensic, environmental, industrial and medical questions, such as in histology.[6]

Modern analytical chemistry is dominated by instrumental analysis. Many analytical chemists focus on a single type of instrument. Academics tend to either focus on new applications and discoveries or on new methods of analysis. The discovery of a chemical present in blood that increases the risk of cancer would be a discovery that an analytical chemist might be involved in. An effort to develop a new method might involve the use of a tunable laser to increase the specificity and sensitivity of a spectrometric method. Many methods, once developed, are kept purposely static so that data can be compared over long periods of time. This is particularly true in industrial quality assurance (QA), forensic and environmental applications. Analytical chemistry plays an increasingly important role in the pharmaceutical industry where, aside from QA, it is used in discovery of new drug candidates and in clinical applications where understanding the interactions between the drug and the patient are critical.

Classical methods

Although modern analytical chemistry is dominated by sophisticated instrumentation, the roots of analytical chemistry and some of the principles used in modern instruments are from traditional techniques many of which are still used today. These techniques also tend to form the backbone of most undergraduate analytical chemistry educational labs.

The presence of copper in this qualitative analysis is indicated by the bluish-green color of the flame.

Qualitative analysis

A qualitative analysis determines the presence or absence of a particular compound, but not the mass or concentration. That is, it is not related to quantity.

Chemical tests

There are numerous qualitative chemical tests, for example, the acid test for gold and the Kastle-Meyer test for the presence of blood.

Flame test

Inorganic qualitative analysis generally refers to a systematic scheme to confirm the presence of certain, usually aqueous, ions or elements by performing a series of reactions that eliminate ranges of possibilities and then confirms suspected ions with a confirming test. Sometimes small carbon containing ions are included in such schemes. With modern instrumentation these tests are rarely used but can be useful for educational purposes and in field work or other situations where access to state-of-the-art instruments are not available or expedient.

Gravimetric analysis

Gravimetric analysis involves determining the amount of material present by weighing the sample before and/or after some transformation. A common example used in undergraduate education is the determination of the amount of water in a hydrate by heating the sample to remove the water such that the difference in weight is due to the loss of water.

Volumetric analysis

Titration involves the addition of a reactant to a solution being analyzed until some equivalence point is reached. Often the amount of material in the solution being analyzed may be determined. Most familiar to those who have taken college chemistry is the acid-base titration involving a color changing indicator. There are many other types of titrations, for example potentiometric titrations. These titrations may use different types of indicators to reach some equivalence point.

Instrumental methods

Spectroscopy

Spectroscopy measures the interaction of the molecules with electromagnetic radiation. Spectroscopy consists of many different applications such as atomic absorption spectroscopy, atomic emission spectroscopy, ultraviolet-visible spectroscopy, x-ray fluorescence spectroscopy, infrared spectroscopy, Raman spectroscopy, dual polarisation interferometry, nuclear magnetic resonance spectroscopy, photoemission spectroscopy, Mössbauer spectroscopy and so on.

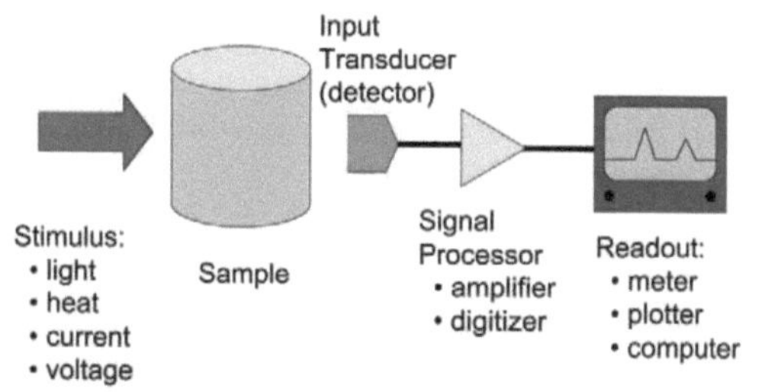

Block diagram of an analytical instrument showing the stimulus and measurement of response

Mass spectrometry

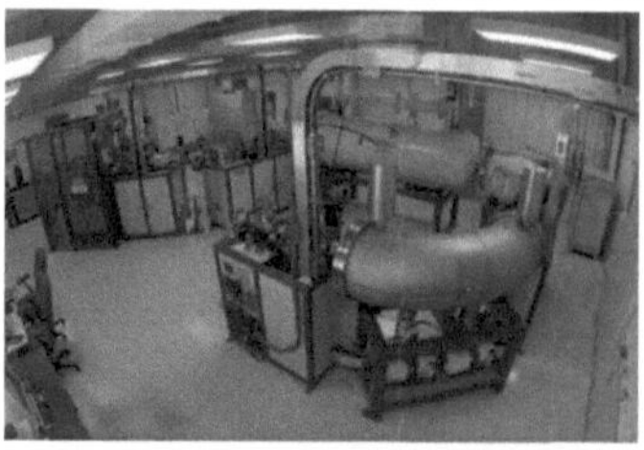

An accelerator mass spectrometer used for radiocarbon dating and other analysis.

Mass spectrometry measures mass-to-charge ratio of molecules using electric and magnetic fields. There are several ionization methods: electron impact, chemical ionization, electrospray, fast atom bombardment, matrix assisted laser desorption ionization, and others. Also, mass spectrometry is categorized by approaches of mass analyzers: magnetic-sector, quadrupole mass analyzer, quadrupole ion trap, time-of-flight, Fourier transform ion cyclotron resonance, and so on.

Electrochemical analysis

Electroanalytical methods measure the potential (volts) and/or current (amps) in an electrochemical cell containing the analyte.[7] [8] These methods can be categorized according to which aspects of the cell are controlled and which are measured. The three main categories are potentiometry (the difference in electrode potentials is measured), coulometry (the cell's current is measured over time), and voltammetry (the cell's current is measured while actively altering the cell's potential).

Thermal analysis

Calorimetry and thermogravimetric analysis measure the interaction of a material and heat.

Separation

Separation processes are used to decrease the complexity of material mixtures. Chromatography and electrophoresis are representative of this field.

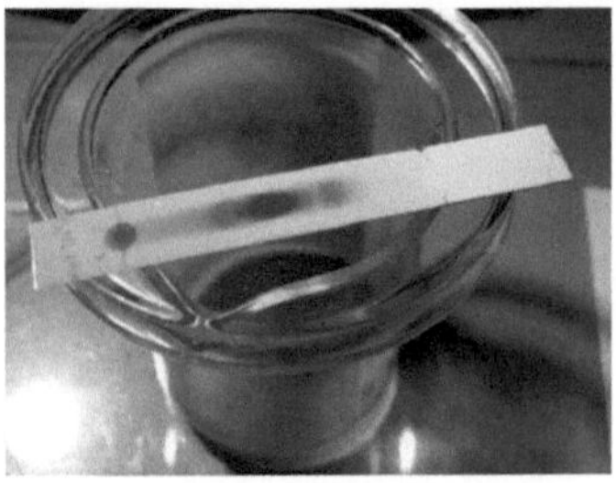

Separation of black ink on a thin layer chromatography plate.

Hybrid techniques

Combinations of the above techniques produce a "hybrid" or "hyphenated" technique.[9] [10] [11] [12] [13] Several examples are in popular use today and new hybrid techniques are under development. For example, gas chromatography-mass spectrometry, gas chromatography-infrared spectroscopy, liquid chromatography-mass spectrometry, liquid chromatography-NMR spectroscopy. liquid chromagraphy-infrared spectroscopy and capillary electrophoresis-mass spectrometry.

Hyphenated separation techniques refers to a combination of two (or more) techniques to detect and separate chemicals from solutions. Most often the other technique is some form of chromatography. Hyphenated techniques are widely used in chemistry and biochemistry. A slash is sometimes used instead of hyphen, especially if the name of one of the methods contains a hyphen itself.

Microscopy

The visualization of single molecules, single cells, biological tissues and nanomaterials is an important and attractive approach in analytical science. Also, hybridization with other traditional analytical tools is revolutionizing analytical science. Microscopy can be categorized into three different fields: optical microscopy, electron microscopy, and scanning probe microscopy. Recently, this field is rapidly progressing because of the rapid development of the computer and camera industries.

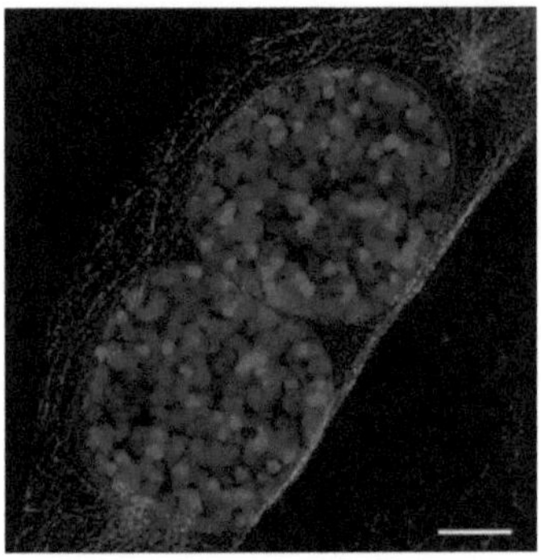

Fluorescence microscope image of two mouse cell nuclei in prophase (scale bar is 5 μm).[14]

Lab-on-a-chip

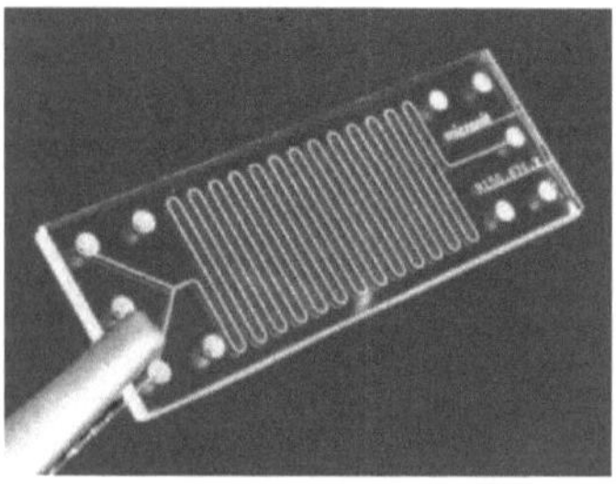

A glass microreactor

Devices that integrate (multiple) laboratory functions on a single chip of only millimeters to a few square centimeters in size and that are capable of handling extremely small fluid volumes down to less than pico liters.

Standards

Standard curve

A general method for analysis of concentration involves the creation of a calibration curve. This allows for determination of the amount of a chemical in a material by comparing the results of unknown sample to those of a series known standards. If the concentration of element or compound in a sample is too high for the detection range of the technique, it can simply be diluted in a pure solvent. If the amount in the sample is below an instrument's range of measurement, the method of addition can be used. In this method a known quantity of the element or compound under study is added, and the difference between the concentration added, and the concentration observed is the amount actually in the sample.

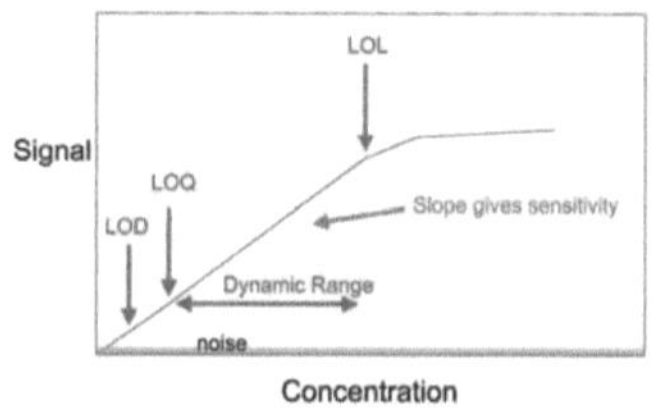

A calibration curve plot showing limit of detection (LOD), limit of quantification (LOQ), dynamic range, and limit of linearity (LOL).

Internal standards

Sometimes an internal standard is added at a known concentration directly to an analytical sample to aid in quantitation. The amount of analyte present is then determined relative to the internal standard as a calibrant.

Standard addition

The method of standard addition is used in instrumental analysis to determine concentration of a substance (analyte) in an unknown sample by comparison to a set of samples of known concentration, similar to using a calibration curve. Standard addition can be applied to most analytical techniques and is used instead of a calibration curve to solve the matrix effect problem.

Signals and noise

One of the most important components of analytical chemistry is maximizing the desired signal while minimizing the associated noise.[15] The analytical figure of merit is known as the signal-to-noise ratio (S/N or SNR).

Noise can arise from environmental factors as well as from fundamental physical processes.

Thermal noise

Thermal noise results from the motion of charge carriers (usually electrons) in an electrical circuit generated by their thermal motion. Thermal noise is white noise meaning that the power spectral density is constant throughout the frequency spectrum.

The root mean square value of the thermal noise in a resistor is given by[15]

where k_B is Boltzmann's constant, T is the temperature, R is the resistance, and is the bandwidth of the frequency .

Shot noise

Shot noise is a type of electronic noise that occurs when the finite number of particles (such as electrons in an electronic circuit or photons in an optical device) is small enough to give rise to statistical fluctuations in a signal.

Shot noise is a Poisson process and the charge carriers that make up the current follow a Poisson distribution. The root mean square current fluctuation is given by[15]

where e is the elementary charge and I is the average current. Shot noise is white noise.

Flicker noise

Flicker noise is electronic noise with a $1/f$ frequency spectrum; as f increases, the noise decreases. Flicker noise arises from a variety of sources, such as impurities in a conductive channel, generation and recombination noise in a transistor due to base current, and so on. This noise can be avoided by modulation of the signal at a higher frequency, for example through the use of a lock-in amplifier.

Environmental noise

Environmental noise arises from the surroundings of the analytical instrument. Sources of electromagnetic noise are power lines, radio and television stations, wireless devices and electric motors. Many of these noise sources are narrow bandwidth and therefore can be avoided. Temperature and vibration isolation may be required for some instruments.

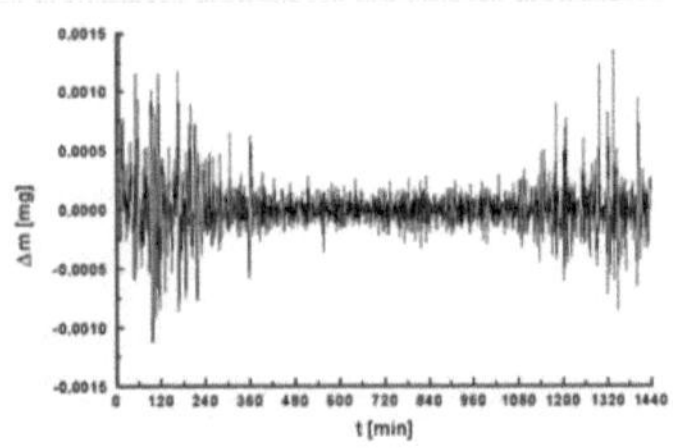

Noise in a thermogravimetric analysis; lower noise in the middle of the plot results from less human activity (and environmental noise) at night.

Noise reduction

Noise reduction can be accomplished either in hardware or software. Examples of hardware noise reduction are the use of shielded cable, analog filtering, and signal modulation. Examples of software noise reduction are digital filtering, ensemble average, boxcar average, and correlation methods.[15]

Applications

Analytical chemistry research is largely driven by performance (sensitivity, selectivity, robustness, linear range, accuracy, precision, and speed), and cost (purchase, operation, training, time, and space). Among the main branches of contemporary analytical atomic spectrometry, the most widespread and universal are optical and mass spectrometry.[16] In the direct elemental analysis of solid samples, the new leaders are laser-induced breakdown and laser ablation mass spectrometry, and the related techniques with transfer of the laser ablation products into inductively coupled plasma. Advances in design of diode lasers and optical parametric oscillators promote developments in fluorescence and ionization spectrometry and also in absorption techniques where uses of optical cavities for increased effective absorption pathlength are expected to expand. Steady progress and growth in applications of plasma- and laser-based methods are noticeable. An interest towards the absolute (standardless) analysis has revived, particularly in the emission spectrometry.

A lot of effort is put in shrinking the analysis techniques to chip size. Although there are few examples of such systems competitive with traditional analysis techniques, potential advantages include size/portability, speed, and cost. (micro Total Analysis System (µTAS) or Lab-on-a-chip). Microscale chemistry reduces the amounts of chemicals used.

Much effort is also put into analyzing biological systems. Examples of rapidly expanding fields in this area are:

- Genomics - DNA sequencing and its related research. Genetic fingerprinting and DNA microarray are very popular tools and research fields.
- Proteomics - the analysis of protein concentrations and modifications, especially in response to various stressors, at various developmental stages, or in various parts of the body.
- Metabolomics - similar to proteomics, but dealing with metabolites.
- Transcriptomics - mRNA and its associated field
- Lipidomics - lipids and its associated field
- Peptidomics - peptides and its associated field
- Metalomics - similar to proteomics and metabolomics, but dealing with metal concentrations and especially with their binding to proteins and other molecules.

Analytical chemistry has played critical roles in the understanding of basic science to a variety of practical applications, such as biomedical applications, environmental monitoring, quality control of industrial manufacturing, forensic science and so on.

The recent developments of computer automation and information technologies have innervated analytical chemistry to initiate a number of new biological fields. For example, automated DNA sequencing machines were the basis to complete human genome projects leading to the birth of genomics. Protein identification and peptide sequencing by mass spectrometry opened a new field of proteomics. Furthermore, a number of ~omics based on analytical chemistry have become important areas in modern biology.

Also, analytical chemistry has been an indispensable area in the development of nanotechnology. Surface characterization instruments, electron microscopes and scanning probe microscopes enables scientists to visualize atomic structures with chemical characterizations.

Among active contemporary analytical chemistry research fields, micro total analysis system is considered as a great promise of revolutionary technology. In this approach, integrated and miniaturized analytical systems are being developed to control and analyze single cells and single molecules. This cutting-edge technology has a promising potential of leading a new revolution in science as integrated circuits did in computer developments.

See also

- AutoAnalyzer
- List of chemical analysis methods
- List of materials analysis methods
- Important publications in analytical chemistry
- Measurement Science and Technology - journal
- Sensory analysis - in the field of Food science
- Virtual instrumentation
- Water chemistry analysis
- Working range
- Metrology
- Measurement uncertainty
- Microanalysis

References

[1] Holler, F. James; Skoog, Douglas A.; West, Donald M. (1996). *Fundamentals of analytical chemistry*. Philadelphia: Saunders College Pub. ISBN 0-03-005938-0.

[2] Nieman, Timothy A.; Skoog, Douglas A.; Holler, F. James (1998). *Principles of instrumental analysis*. Pacific Grove, CA: Brooks/Cole. ISBN 0-03-002078-6.

[3] ANALYTICAL SCIENCES 2001, VOL.17 SUPPLEMENT (http://wwwsoc.nii.ac.jp/jsac/analsci/ICAS2001/pdfs/0500/0571_3i05n.pdf), Basic Education in Analytical Chemistry

[4] Talanta Volume 51, Issue 5, p921-933 (http://dx.doi.org/10.1016/S0039-9140(99)00358-6), Review of analytical next term measurements facilitated by drop formation technology

[5] TrAC Trends in Analytical Chemistry Volume 21, Issues 9-10, Pages 547-557 (http://dx.doi.org/10.1016/S0165-9936(02)00806-3), History of gas chromatography

[6] Talanta, Volume 36, Issues 1-2, January–February 1989, Pages 1-9 (http://dx.doi.org/10.1016/0039-9140(89)80077-3) History of analytical chemistry in the U.S.A.

[7] Bard, A.J.; Faulkner, L.R. **Electrochemical Methods: Fundamentals and Applications.** New York: John Wiley & Sons, 2nd Edition, **2000**.

[8] Skoog, D.A.; West, D.M.; Holler, F.J. **Fundamentals of Analytical Chemistry** New York: Saunders College Publishing, 5th Edition, **1988**.

[9] Wilkins CL (1983). "Hyphenated techniques for analysis of complex organic mixtures". *Science* **222** (4621): 291–6. doi:10.1126/science.6353577. PMID 6353577.

[10] Holt RM, Newman MJ, Pullen FS, Richards DS, Swanson AG (1997). "High-performance liquid chromatography/NMR spectrometry/mass spectrometry: further advances in hyphenated technology". *Journal of mass spectrometry : JMS* **32** (1): 64–70. doi:10.1002/(SICI)1096-9888(199701)32:1<64::AID-JMS450>3.0.CO;2-7. PMID 9008869.

[11] Ellis LA, Roberts DJ (1997). "Chromatographic and hyphenated methods for elemental speciation analysis in environmental media". *Journal of chromatography. A* **774** (1–2): 3–19. doi:10.1016/S0021-9673(97)00325-7. PMID 9253184.

[12] Guetens G, De Boeck G, Wood M, Maes RA, Eggermont AA, Highley MS, van Oosterom AT, de Bruijn EA, Tjaden UR (2002). "Hyphenated techniques in anticancer drug monitoring. I. Capillary gas chromatography-mass spectrometry". *Journal of chromatography. A* **976** (1–2): 229–38. doi:10.1016/S0021-9673(02)01228-1. PMID 12462614.

[13] Guetens G, De Boeck G, Highley MS, Wood M, Maes RA, Eggermont AA, Hanauske A, de Bruijn EA, Tjaden UR (2002). "Hyphenated techniques in anticancer drug monitoring. II. Liquid chromatography-mass spectrometry and capillary electrophoresis-mass spectrometry". *Journal of chromatography. A* **976** (1–2): 239–47. doi:10.1016/S0021-9673(02)01227-X. PMID 12462615.

[14] Schermelleh, L.; Carlton, P. M.; Haase, S.; Shao, L.; Winoto, L.; Kner, P.; Burke, B.; Cardoso, M. C. et al. (2008). "Subdiffraction Multicolor Imaging of the Nuclear Periphery with 3D Structured Illumination Microscopy". *Science* **320** (5881): 1332. doi:10.1126/science.1156947. PMC 2916659. PMID 18535242.

[15] Crouch, Stanley; Skoog, Douglas A. (2007). *Principles of instrumental analysis*. Australia: Thomson Brooks/Cole. ISBN 0-495-01201-7.

[16] Bol'shakov, Aleksandr A; Ganeev, Aleksandr A; Nemets, Valerii M (2006). "Prospects in analytical atomic spectrometry". *Russian Chemical Reviews* **75** (4): 289. arXiv:physics/0607078. doi:10.1070/RC2006v075n04ABEH001174.

External links

- Annual Review of Analytical Chemistry (http://arjournals.annualreviews.org/loi/anchem)
- American Chemical Society: Division of Analytical Chemistry (http://www.analyticalsciences.org/)
- Royal Society of Chemistry: Analytical Gateway (http://rsc.org/Gateway/Subject/Analytical/)

Physics

Physics (from Ancient Greek: φύσις *physis* "nature") is a natural science that involves the study of matter[1] and its motion through spacetime, along with related concepts such as energy and force.[2] More broadly, it is the general analysis of nature, conducted in order to understand how the universe behaves.[3] [4] [5]

Physics is one of the oldest academic disciplines, perhaps the oldest through its inclusion of astronomy.[6] Over the last two millennia, physics was a part of natural philosophy along with chemistry, certain branches of mathematics, and biology, but during the Scientific Revolution in the 16th century, the natural sciences emerged as unique research programs in their own right.[7] Certain research areas are interdisciplinary, such as biophysics and quantum chemistry, which means that the boundaries of physics are not rigidly defined. In the nineteenth and twentieth centuries physicalism emerged as a major unifying feature of the philosophy of science as physics provides fundamental explanations for every observed natural phenomenon. New ideas in physics often explain the fundamental mechanisms of other sciences, while opening to new research areas in mathematics and philosophy.

Physics is also significant and influential through advances in its understanding that have translated into new technologies. For example, advances in the understanding of electromagnetism or nuclear physics led directly to the development of new products which have dramatically transformed modern-day society, such as television, computers, domestic appliances, and nuclear weapons; advances in thermodynamics led to the development of industrialization; and advances in mechanics inspired the development of calculus.

Scope and aims

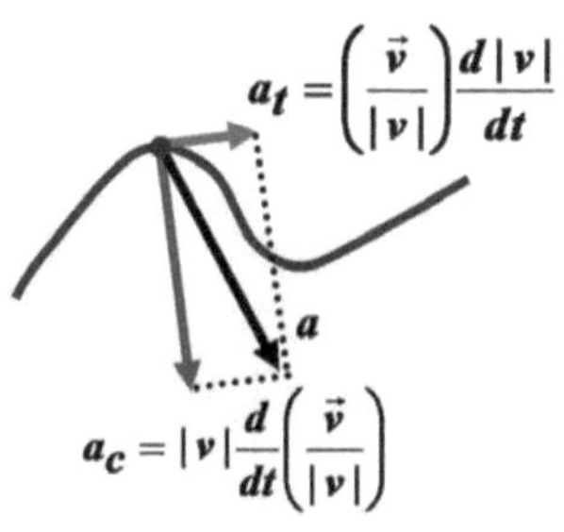

Physics involves modeling the natural world with theory, usually quantitative. Here, the path of a particle is modeled with the mathematics of calculus to explain its behavior: the purview of the branch of physics known as mechanics.

Physics covers a wide range of phenomena, from elementary particles (such as quarks, neutrinos and electrons) to the largest superclusters of galaxies. Included in these phenomena are the most basic objects composing all other things. Therefore physics is sometimes called the "fundamental science".[8] Physics aims to describe the various phenomena that occur in nature in terms of simpler phenomena. Thus, physics aims to both connect the things observable to humans to root causes, and then connect these causes together.

For example, the ancient Chinese observed that certain rocks (lodestone) were attracted to one another by some invisible force. This effect was later called magnetism, and was first rigorously studied in the 17th century. A little earlier than the Chinese, the ancient Greeks knew of other objects such as amber, that when rubbed with fur would cause a similar invisible attraction between the two. This was also first studied rigorously in the 17th century, and came to be called electricity. Thus, physics had come to understand two observations of nature in terms of some root cause (electricity and magnetism). However, further work in the 19th century revealed that these two forces were just two different aspects of one force – electromagnetism. This process of "unifying" forces continues today, and electromagnetism and the weak nuclear force are now considered to be two aspects of the electroweak interaction. Physics hopes to find an ultimate reason (Theory of Everything) for why nature is as it is (see section *Current research* below for more information).

Scientific method

Physicists use a scientific method to test the validity of a physical theory, using a methodical approach to compare the implications of the theory in question with the associated conclusions drawn from experiments and observations conducted to test it. Experiments and observations are to be collected and matched with the predictions and hypotheses made by a theory, thus aiding in the determination or the validity/invalidity of the theory.

Theories which are very well supported by data and have never failed any competent empirical test are often called scientific laws, or natural laws. Of course, all theories, including those called scientific laws, can always be replaced by more accurate, generalized statements if a disagreement of theory with observed data is ever found.[9]

Theory and experiment

Theorists seek to develop mathematical models that both agree with existing experiments and successfully predict future results, while experimentalists devise and perform experiments to test theoretical predictions and explore new phenomena. Although theory and experiment are developed separately, they are strongly dependent upon each other. Progress in physics frequently comes about when experimentalists make a discovery that existing theories cannot explain, or when new theories generate experimentally testable predictions, which inspire new experiments.

The astronaut and Earth are both in free-fall

Physicists who work at the interplay of theory and experiment are called phenomenologists. Phenomenologists look at the complex phenomena observed in experiment and work to relate them to fundamental theory.

Theoretical physics has historically taken inspiration from philosophy; electromagnetism was unified this way.[10] Beyond the known universe, the field of theoretical physics also deals with hypothetical issues,[11] such as parallel universes, a multiverse, and higher dimensions. Theorists invoke these ideas in hopes of solving particular problems with existing theories. They then explore the consequences of these ideas and work toward making testable predictions.

Lightning is an electric current

Experimental physics informs, and is informed by, engineering and technology. Experimental physicists involved in basic research design and perform experiments with equipment such as particle accelerators and lasers, whereas those involved in applied research often work in industry, developing technologies such as magnetic resonance imaging (MRI) and transistors. Feynman has noted that experimentalists may seek areas which are not well explored by theorists.[12]

Relation to other sciences and to mathematics

This parabola-shaped lava flow illustrates Galileo's law of falling bodies as well as blackbody radiation – the temperature is discernible from the color of the blackbody.

In the *Assayer* (1622), Galileo noted that mathematics is the language in which Nature expresses its laws.[13] Most experimental results in physics are numerical measurements, and theories in physics use mathematics to give numerical results to match these measurements.

Physics relies upon mathematics to provide the logical framework in which physical laws may be precisely formulated and predictions quantified. Whenever analytic solutions of equations are not feasible, numerical analysis and simulations may be utilized. Thus, scientific computation is an integral part of physics, and the field of computational physics is an active area of research.

A key difference between physics and mathematics is that since physics is ultimately concerned with descriptions of the material world, it tests its theories by comparing the predictions of its theories with data procured from observations and experimentation, whereas mathematics is concerned with abstract patterns, not limited by those observed in the real world. The distinction, however, is not always clear-cut. There is a large area of research intermediate between physics and mathematics, known as mathematical physics.

Physics is also intimately related to many other sciences, as well as applied fields like engineering and medicine. The principles of physics find applications throughout the other natural sciences as some phenomena studied in physics, such as the conservation of energy, are common to *all* material systems. Other phenomena, such as superconductivity, stem from these laws, but are not laws themselves because they only appear in some systems.

Physics is often said to be the "fundamental science" (chemistry is sometimes included), because each of the other disciplines (biology, chemistry, geology, material science, engineering, medicine etc.) deals with particular types of material systems that obey the laws of physics.[8] For example, chemistry is the science of collections of matter (such as gases and liquids formed of atoms and molecules) and the processes known as chemical reactions that result in the change of chemical substances.

The structure, reactivity, and properties of a chemical compound are determined by the properties of the underlying molecules, which may be well-described by areas of physics such as quantum mechanics, or quantum chemistry, thermodynamics, and electromagnetism.

Philosophical implications

Physics in many ways stems from ancient Greek philosophy. From Thales' first attempt to characterize matter, to Democritus' deduction that matter ought to reduce to an invariant state, the Ptolemaic astronomy of a crystalline firmament, and Aristotle's book *Physics*, different Greek philosophers advanced their own theories of nature. Well into the 18th century, physics was known as natural philosophy.

By the 19th century physics was realized as a discipline distinct from philosophy and the other sciences. Physics, as with the rest of science, relies on philosophy of science to give an adequate description of the scientific method.[14] The scientific method employs a priori reasoning as well as a posteriori reasoning and the use of Bayesian inference to measure the validity of a given theory.[15]

The development of physics has answered many questions of early philosophers, but has also raised new questions. Study of the philosophical issues surrounding physics, the philosophy of physics, involves issues such as the nature of space and time, determinism, and metaphysical outlooks such as empiricism, naturalism and realism.[16]

Many physicists have written about the philosophical implications of their work, for instance Laplace, who championed causal determinism,[17] and Erwin Schrödinger, who wrote on quantum mechanics.[18] The mathematical physicist Roger Penrose has been called a Platonist by Stephen Hawking,[19] a view Penrose discusses in his book, *The Road to Reality*.[20] Hawking refers to himself as an "unashamed reductionist" and takes issue with Penrose's views.[21]

History

Isaac Newton (1643–1727)

Since antiquity, people have tried to understand the behavior of the natural world. One great mystery was the predictable behavior of celestial objects such as the Sun and the Moon. Several theories were proposed, the majority of which were disproved.

The philosopher Thales (ca. 624–546 BC) first refused to accept various supernatural, religious or mythological explanations for natural phenomena, proclaiming that every event had a natural cause. Early physical theories were largely couched in philosophical terms, and never verified by systematic experimental testing as is popular today. Many of the commonly accepted works of Ptolemy and Aristotle are not always found to match everyday observations.

Even so, many ancient philosophers and astronomers gave correct descriptions in atomism and astronomy. Leucippus (first half of 5th century BC) first proposed atomism, while Archimedes derived many correct quantitative descriptions of mechanics, statics and hydrostatics, including an explanation for the principle of the lever. The Middle Ages saw the emergence of an experimental physics taking shape among medieval Muslim physicists, the most famous being Alhazen, followed by modern physics largely taking shape among early modern European physicists, the most famous being Sir Isaac Newton, who built on the works of Galileo Galilei and Johannes Kepler. In the 20th century, the work of Albert Einstein marked a new direction in physics that continues to the present day.

Core theories

While physics deals with a wide variety of systems, certain theories are used by all physicists. Each of these theories were experimentally tested numerous times and found correct as an approximation of nature (within a certain domain of validity). For instance, the theory of classical mechanics accurately describes the motion of objects, provided they are much larger than atoms and moving at much less than the speed of light. These theories continue to be areas of active research, and a remarkable aspect of classical mechanics known as chaos was discovered in the 20th century, three centuries after the original formulation of classical mechanics by Isaac Newton (1642–1727).

These central theories are important tools for research into more specialized topics, and any physicist, regardless of his or her specialization, is expected to be literate in them. These include classical mechanics, quantum mechanics, thermodynamics and statistical mechanics, electromagnetism, and special relativity.

Research fields

Contemporary research in physics can be broadly divided into condensed matter physics; atomic, molecular, and optical physics; particle physics; astrophysics; geophysics and biophysics. Some physics departments also support research in Physics education.

Since the twentieth century, the individual fields of physics have become increasingly specialized, and today most physicists work in a single field for their entire careers. "Universalists" such as Albert Einstein (1879–1955) and Lev Landau (1908–1968), who worked in multiple fields of physics, are now very rare.[22]

Condensed matter

Condensed matter physics is the field of physics that deals with the macroscopic physical properties of matter. In particular, it is concerned with the "condensed" phases that appear whenever the number of constituents in a system is extremely large and the interactions between the constituents are strong.

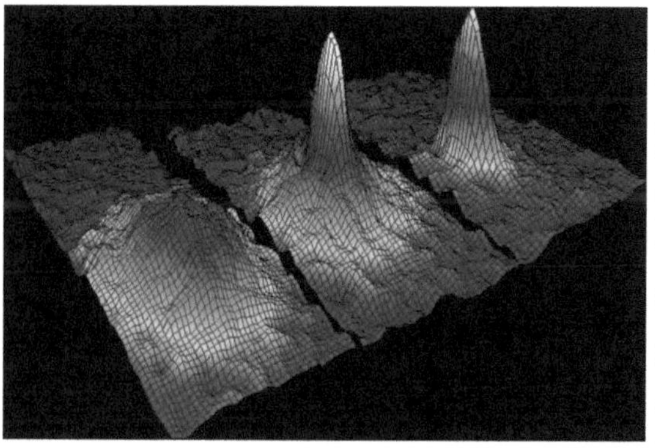

Velocity-distribution data of a gas of rubidium atoms, confirming the discovery of a new phase of matter, the Bose–Einstein condensate

The most familiar examples of condensed phases are solids and liquids, which arise from the bonding and electromagnetic force between atoms. More exotic condensed phases include the superfluid and the Bose–Einstein condensate found in certain atomic systems at very low temperature, the superconducting phase exhibited by conduction electrons in certain materials, and the ferromagnetic and antiferromagnetic phases of spins on atomic lattices.

Condensed matter physics is by far the largest field of contemporary physics. Historically, condensed matter physics grew out of solid-state physics, which is now considered one of its main subfields. The term *condensed matter physics* was apparently coined by Philip Anderson when he renamed his research group — previously *solid-state theory* — in 1967.

In 1978, the Division of Solid State Physics at the American Physical Society was renamed as the Division of Condensed Matter Physics.[23] Condensed matter physics has a large overlap with chemistry, materials science, nanotechnology and engineering.

Atomic, molecular, and optical physics

Atomic, molecular, and optical physics (AMO) is the study of matter-matter and light-matter interactions on the scale of single atoms or structures containing a few atoms. The three areas are grouped together because of their interrelationships, the similarity of methods used, and the commonality of the energy scales that are relevant. All three areas include both classical and quantum treatments; they can treat their subject from a microscopic view (in contrast to a macroscopic view).

Atomic physics studies the electron shells of atoms. Current research focuses on activities in quantum control, cooling and trapping of atoms and ions, low-temperature collision dynamics, the collective behavior of atoms in weakly interacting gases (Bose–Einstein Condensates and dilute Fermi degenerate systems), precision measurements of fundamental constants, and the effects of electron correlation on structure and dynamics. Atomic physics is influenced by the nucleus (see, e.g., hyperfine splitting), but intra-nuclear phenomenon such as fission and fusion are considered part of high energy physics.

Molecular physics focuses on multi-atomic structures and their internal and external interactions with matter and light. Optical physics is distinct from optics in that it tends to focus not on the control of classical light fields by macroscopic objects, but on the fundamental properties of optical fields and their interactions with matter in the microscopic realm.

High energy/particle physics

Particle physics is the study of the elementary constituents of matter and energy, and the interactions between them. It may also be called "high energy physics", because many elementary particles do not occur naturally, but are created only during high energy collisions of other particles, as can be detected in particle accelerators.

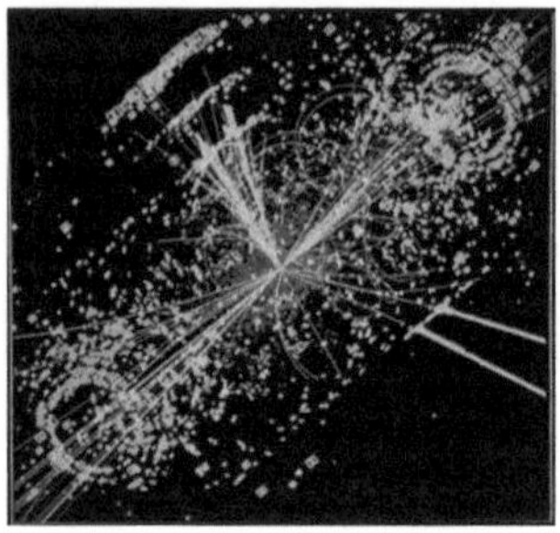

A simulated event in the CMS detector of the Large Hadron Collider, featuring a possible appearance of the Higgs boson.

Currently, the interactions of elementary particles are described by the Standard Model. The model accounts for the 12 known particles of matter (quarks and leptons) that interact via the strong, weak, and electromagnetic fundamental forces. Dynamics are described in terms of matter particles exchanging gauge bosons (gluons, W and Z bosons, and photons, respectively). The Standard Model also predicts a particle known as the Higgs boson, the existence of which has not yet been verified; as of 2010, searches for it are underway in the Tevatron at Fermilab and in the Large Hadron Collider at CERN.

Astrophysics

The deepest visible-light image of the universe, the Hubble Ultra Deep Field

Astrophysics and astronomy are the application of the theories and methods of physics to the study of stellar structure, stellar evolution, the origin of the solar system, and related problems of cosmology. Because astrophysics is a broad subject, astrophysicists typically apply many disciplines of physics, including mechanics, electromagnetism, statistical mechanics, thermodynamics, quantum mechanics, relativity, nuclear and particle physics, and atomic and molecular physics.

The discovery by Karl Jansky in 1931 that radio signals were emitted by celestial bodies initiated the science of radio astronomy. Most recently, the frontiers of astronomy have been expanded by space exploration. Perturbations and interference from the earth's atmosphere make space-based observations necessary for infrared, ultraviolet, gamma-ray, and X-ray astronomy.

Physical cosmology is the study of the formation and evolution of the universe on its largest scales. Albert Einstein's theory of relativity plays a central role in all modern cosmological theories. In the early 20th century, Hubble's discovery that the universe was expanding, as shown by the Hubble diagram, prompted rival explanations known as the steady state universe and the Big Bang.

The Big Bang was confirmed by the success of Big Bang nucleosynthesis and the discovery of the cosmic microwave background in 1964. The Big Bang model rests on two theoretical pillars: Albert Einstein's general relativity and the cosmological principle. Cosmologists have recently established the ΛCDM model of the evolution of the universe, which includes cosmic inflation, dark energy and dark matter.

Numerous possibilities and discoveries are anticipated to emerge from new data from the Fermi Gamma-ray Space Telescope over the upcoming decade and vastly revise or clarify existing models of the Universe.[24] [25] In particular, the potential for a tremendous discovery surrounding dark matter is possible over the next several years.[26] Fermi will search for evidence that dark matter is composed of weakly interacting massive particles, complementing similar experiments with the Large Hadron Collider and other underground detectors.

IBEX is already yielding new astrophysical discoveries: "No one knows what is creating the ENA (energetic neutral atoms) ribbon" along the termination shock of the solar wind, "but everyone agrees that it means the textbook picture of the heliosphere — in which the solar system's enveloping pocket filled with the solar wind's charged particles is plowing through the onrushing 'galactic wind' of the interstellar medium in the shape of a comet — is wrong."[27]

Fundamental physics

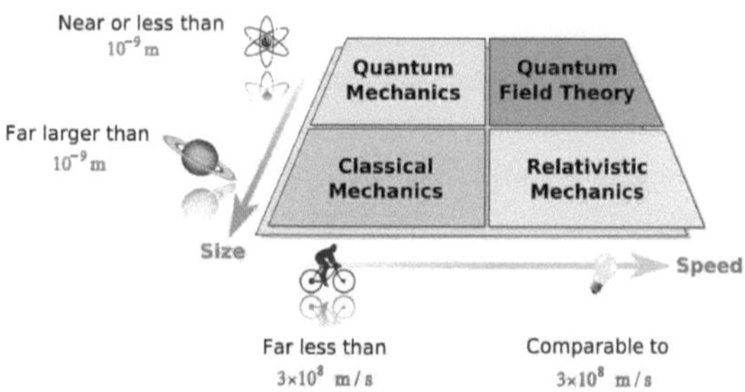

The basic domains of physics

While physics aims to discover universal laws, its theories lie in explicit domains of applicability. Loosely speaking, the laws of classical physics accurately describe systems whose important length scales are greater than the atomic scale and whose motions are much slower than the speed of light. Outside of this domain, observations do not match their predictions. Albert Einstein contributed the framework of special relativity, which replaced notions of absolute time and space with spacetime and allowed an accurate description of systems whose components have speeds approaching the speed of light. Max Planck, Erwin Schrödinger, and others introduced quantum mechanics, a probabilistic notion of particles and interactions that allowed an accurate description of atomic and subatomic scales. Later, quantum field theory unified quantum mechanics and special relativity. General relativity allowed for a dynamical, curved spacetime, with which highly massive systems and the large-scale structure of the universe can be well described. General relativity has not yet been unified with the other fundamental descriptions; several candidates theories of quantum gravity are being developed.

Application and influence

Archimedes' screw uses simple machines to lift liquids.

Applied physics is a general term for physics research which is intended for a particular use. An applied physics curriculum usually contains a few classes in an applied discipline, like geology or electrical engineering. It usually differs from engineering in that an applied physicist may not be designing something in particular, but rather is using physics or conducting physics research with the aim of developing new technologies or solving a problem.

The approach is similar to that of applied mathematics. Applied physicists can also be interested in the use of physics for scientific research. For instance, people working on accelerator physics might seek to build better particle detectors for research in theoretical physics.

Physics is used heavily in engineering. For example, Statics, a subfield of mechanics, is used in the building of bridges and other structures. The understanding and use of acoustics results in better concert halls; similarly, the use of optics creates better optical devices. An understanding of physics makes for more realistic flight simulators, video games, and movies, and is often critical in forensic investigations.

With the standard consensus that the laws of physics are universal and do not change with time, physics can be used to study things that would ordinarily be mired in uncertainty. For example, in the study of the origin of the Earth, one can reasonably model Earth's mass, temperature, and rate of rotation, over time. It also allows for simulations in engineering which drastically speed up the development of a new technology.

But there is also considerable interdisciplinarity in the physicist's methods, and so many other important fields are influenced by physics: e.g. presently the fields of econophysics plays an important role, as well as sociophysics.

Current research

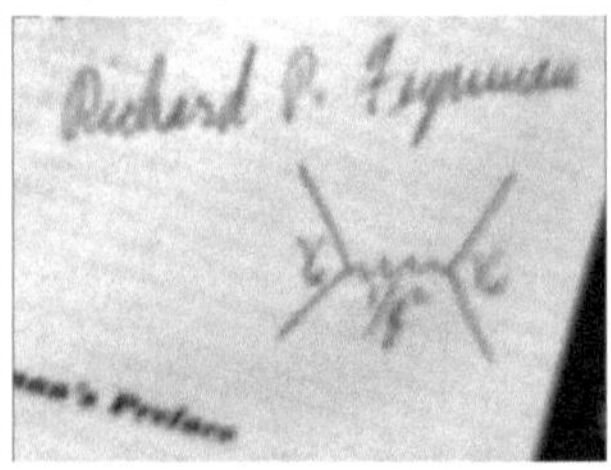

Feynman diagram signed by R. P. Feynman

Research in physics is continually progressing on a large number of fronts.

In condensed matter physics, an important unsolved theoretical problem is that of high-temperature superconductivity. Many condensed matter experiments are aiming to fabricate workable spintronics and quantum computers.

In particle physics, the first pieces of experimental evidence for physics beyond the Standard Model have begun to appear. Foremost among these are indications that neutrinos have non-zero mass. These experimental results appear to have solved the long-standing solar neutrino problem, and the physics of massive neutrinos remains an area of active theoretical and experimental research. Particle accelerators have begun probing energy scales in the TeV range, in which experimentalists are hoping to find evidence for the Higgs boson and supersymmetric particles.[28]

Theoretical attempts to unify quantum mechanics and general relativity into a single theory of quantum gravity, a program ongoing for over half a century, have not yet been decisively resolved. The current leading candidates are M-theory, superstring theory and loop quantum gravity.

A typical event described by physics: a magnet levitating above a superconductor demonstrates the Meissner effect.

Many astronomical and cosmological phenomena have yet to be satisfactorily explained, including the existence of ultra-high energy cosmic rays, the baryon asymmetry, the acceleration of the universe and the anomalous rotation rates of galaxies.

Although much progress has been made in high-energy, quantum, and astronomical physics, many everyday phenomena involving complexity, chaos, or turbulence are still poorly understood. Complex problems that seem like they could be solved by a clever application of dynamics and mechanics remain unsolved; examples include the formation of sandpiles, nodes in trickling water, the shape of water droplets, mechanisms of surface tension catastrophes, and self-sorting in shaken heterogeneous collections.

These complex phenomena have received growing attention since the 1970s for several reasons, including the availability of modern mathematical methods and computers, which enabled complex systems to be modeled in new ways. Complex physics has become part of increasingly interdisciplinary research, as exemplified by the study of turbulence in aerodynamics and the observation of pattern formation in biological systems. In 1932, Horace Lamb said:[29]

> I am an old man now, and when I die and go to heaven there are two matters on which I hope for enlightenment. One is quantum electrodynamics, and the other is the turbulent motion of fluids. And about the former I am rather optimistic.

See also

General

- Glossary of classical physics
- Index of physics articles
- List of elementary physics formulae
- List of important publications in physics
- List of unsolved problems in physics
- Perfection in physics and chemistry
- Philosophy of physics
- *Physics* (Aristotle) – an early book on physics, which attempted to analyze and define motion from a philosophical point of view
- Timeline of fundamental physics discoveries

Related fields

- Astronomy
- Chemistry
- Engineering
- Mathematics
- Quantum Mechanics
- Science

Interdisciplinary fields incorporating physics

- Nanotechnology
- Biophysics
- Econophysics
- Geophysics
- Neurophysics
- Psychophysics

References

[1] Richard Feynman begins his *Lectures* with the atomic hypothesis, as his most compact statement of all scientific knowledge: "If, in some cataclysm, all of scientific knowledge were to be destroyed, and only one sentence passed on to the next generations ..., what statement would contain the most information in the fewest words? I believe it is ... that *all things are made up of atoms – little particles that move around in perpetual motion, attracting each other when they are a little distance apart, but repelling upon being squeezed into one another. ...*" R.P. Feynman, R.B. Leighton, M. Sands (1963). *The Feynman Lectures on Physics*. **1**. p. I-2. ISBN 0-201-02116-1.

[2] J.C. Maxwell (1878). *Matter and Motion* (http://books.google.com/?id=noRgWP0_UZ8C&printsec=titlepage&dq=matter+and+motion). D. Van Nostrand. p. 9. ISBN 0486668959. . "Physical science is that department of knowledge which relates to the order of nature, or, in other words, to the regular succession of events."

[3] H.D. Young, R.A. Freedman (2004). *University Physics with Modern Physics* (11th ed.). Addison Wesley. p. 2. "Physics is an *experimental* science. Physicists observe the phenomena of nature and try to find patterns and principles that relate these phenomena. These patterns are called physical theories or, when they are very well established and of broad use, physical laws or principles."

[4] S. Holzner (2006). *Physics for Dummies* (http://www.amazon.com/gp/reader/0764554336). Wiley. p. 7. ISBN 0470618418. . "Physics is the study of your world and the world and universe around you."

[5] Note: The term 'universe' is defined as everything that physically exists: the entirety of space and time, all forms of matter, energy and momentum, and the physical laws and constants that govern them. However, the term 'universe' may also be used in slightly different contextual senses, denoting concepts such as the cosmos or the philosophical world.

[6] Evidence exists that the earliest civilizations dating back to beyond 3000 BCE, such as the Sumerians, Ancient Egyptians, and the Indus Valley Civilization, all had a predictive knowledge and a very basic understanding of the motions of the Sun, Moon, and stars.

[7] Francis Bacon's 1620 *Novum Organum* was critical in the development of scientific method.

[8] *The Feynman Lectures on Physics* Volume I. Feynman, Leighton and Sands. ISBN 0-201-02115-3 See Chapter 3 : "The Relation of Physics to Other Sciences" for a general discussion. For the philosophical issue of whether other sciences can be "reduced" to physics, see reductionism and special sciences).

[9] Some principles, such as Newton's laws of motion, are still generally called "laws" even though they are now known to be limiting cases of newer theories. Thus, for example, in Thomas Brody (1993, Luis de la Peña and Peter Hodgson, eds.) *The Philosophy Behind Physics* ISBN 0-387-55914-0, pp 18–24 (Chapter 2), explains the 'epistemic cycle' in which a student of physics discovers that physics is not a finished product but is instead the process of creating [that product].

[10] See, for example, the influence of Kant and Ritter on Oersted.

[11] Concepts which are denoted *hypothetical* can change with time. For example, the atom of nineteenth century physics was denigrated by some, including Ernst Mach's critique of Ludwig Boltzmann's formulation of statistical mechanics. By the end of World War II, the atom was no longer deemed hypothetical.

[12] Feynman, Richard (1965). *The Character of Physical Law*. ISBN 0262560038. p.157: "In fact experimenters have a certain individual character. They ... very often do their experiments in a region in which people know the theorist has not made any guesses. "

[13] "Philosophy is written in that great book which ever lies before our eyes. I mean the universe, but we cannot understand it if we do not first learn the language and grasp the symbols in which it is written. This book is written in the mathematical language, and the symbols are triangles, circles and other geometrical figures, without whose help it is humanly impossible to comprehend a single word of it, and without which one wanders in vain through a dark labyrinth." – Galileo (1623), *The Assayer*, as quoted by G. Toraldo Di Francia (1976), *The Investigation of the Physical World* ISBN 0-521-29925-X p.10

[14] Rosenberg, Alex (2006). *Philosophy of Science*. Routledge. ISBN 0-415-34317-8. See Chapter 1 for a discussion on the necessity of philosophy of science.

[15] Peter Godfrey-Smith (2003), Chapter 14 "Bayesianism and Modern Theories of Evidence" *Theory and Reality: an introduction to the philosophy of science* ISBN 0-226-30063-3

[16] Peter Godfrey-Smith (2003), Chapter 15 "Empiricism, Naturalism, and Scientific Realism?" *Theory and Reality: an introduction to the philosophy of science* ISBN 0-226-30063-3

[17] See Laplace, Pierre Simon, *A Philosophical Essay on Probabilities*, translated from the 6th French edition by Frederick Wilson Truscott and Frederick Lincoln Emory, Dover Publications (New York, 1951)

[18] See "The Interpretation of Quantum Mechanics" Ox Bow Press (1995) ISBN 1-881987-09-4. and "My View of the World" Ox Bow Press (1983) ISBN 0-918024-30-7.

[19] Stephen Hawking and Roger Penrose (1996), *The Nature of Space and Time* ISBN 0-691-05084-8 p.4 "I think that Roger is a Platonist at heart but he must answer for himself."

[20] Roger Penrose, *The Road to Reality* ISBN 0-679-45443-8

[21] Penrose, Roger; Abner Shimony, Nancy Cartwright, Stephen Hawking (1997). *The Large, the Small and the Human Mind*. Cambridge University Press. ISBN 0-521-78572-3.

[22] Yet, universalism is encouraged in the culture of physics. For example, the World Wide Web, which was innovated at CERN by Tim Berners-Lee, was created in service to the computer infrastructure of CERN, and was/is intended for use by physicists worldwide. The same might be said for arXiv.org

[23] "Division of Condensed Matter Physics Governance History" (http://dcmp.bc.edu/page.php?name=governance_history). . Retrieved 13 February 2007.

[24] "NASA - Q&A on the GLAST Mission" (http://www.nasa.gov/mission_pages/GLAST/main/questions_answers.html). *Nasa: Fermi Gamma-ray Space Telescope*. NASA. 28 August 2008. . Retrieved 29 April 2009.

[25] See also Nasa - Fermi Science (http://www.nasa.gov/mission_pages/GLAST/science/index.html) and NASA - Scientists Predict Major Discoveries for GLAST (http://www.nasa.gov/mission_pages/GLAST/science/unidentified_sources.html).

[26] NASA.gov (http://www.nasa.gov/mission_pages/GLAST/science/dark_matter.html)

[27] Richard A. Kerr (16 October 2009). "Tying Up the Solar System With a Ribbon of Charged Particles" (http://www.sciencemag.org/cgi/content/summary/sci;326/5951/350-a?maxtoshow=&HITS=10&hits=10&RESULTFORMAT=&fulltext=IBEX&searchid=1&FIRSTINDEX=0&issue=5951&resourcetype=HWCIT). *Science* **326** (5951): pp. 350–351. . Retrieved 27 November 2009.

[28] 584 co-authors "Direct observation of the strange 'b' baryon " Fermilab-Pub-07/196-E, June 12, 2007 http://arxiv.org/abs/0706.1690v2 finds a mass of 5.774 GeV for the

[29] Goldstein, Sydney (1969). "Fluid Mechanics in the First Half of this Century". *Annual Reviews in Fluid Mechanics* **1**: 1–28. Bibcode 1969AnRFM...1....1G. doi:10.1146/annurev.fl.01.010169.000245.

External links

General

- HyperPhysics website (http://hyperphysics.phy-astr.gsu.edu/Hbase/hframe.html) – HyperPhysics, a physics and astronomy mind-map from Georgia State University
- PhysicsCentral (http://www.physicscentral.com/) – Web portal run by the American Physical Society (http://www.aps.org/)
- Physics.org (http://www.physics.org/) – Web portal run by the Institute of Physics (http://www.iop.org/)
- *The Skeptic's Guide to Physics* (http://musr.physics.ubc.ca/~jess/hr/skept/)
- Usenet Physics FAQ (http://math.ucr.edu/home/baez/physics/) – A FAQ compiled by sci.physics and other physics newsgroups
- Website of the Nobel Prize in physics (http://nobelprize.org/nobel_prizes/physics/)
- World of Physics (http://scienceworld.wolfram.com/physics/) – An online encyclopedic dictionary of physics
- *Nature*: Physics (http://www.nature.com/naturephysics)
- Physics (http://physics.aps.org/) announced July 17, 2008 by the American Physical Society
- Physics/Publications (http://www.dmoz.org//Science/Physics/Publications//) at the Open Directory Project
- Physicsworld.com (http://physicsworld.com) - News website from Institute of Physics Publishing (http://publishing.iop.org/)
- Physics Central (http://physlib.com/) - includes articles on astronomy, particle physics, and mathematics.
- The Vega Science Trust (http://www.vega.org.uk/) - science videos, including physics
- Video: Physics "Lightning" Tour with Justin Morgan (http://www.archive.org/details/JustinMorganPhysicsLightningTour/)
- 52-part video course: The Mechanical Universe...and Beyond (http://www.learner.org/resources/series42.html) Note: also available at 01 - Introduction Physics (http://video.google.com/videoplay?docid=) at Google Videos (Adobe Flash video)
- Encyclopedia of Physics (http://www.scholarpedia.org/article/Encyclopedia_of_physics) at Scholarpedia
- de Haas, Paul, "Historic Papers in Physics (20th Century)" (http://home.tiscali.nl/physis/HistoricPaper/)

Organizations

- AIP.org (http://www.aip.org/index.html) – Website of the American Institute of Physics
- APS.org (http://www.aps.org) – Website of the American Physical Society

- IOP.org (http://www.iop.org) – Website of the Institute of Physics
- PlanetPhysics.org (http://planetphysics.org/)
- Royal Society (http://www.royalsoc.ac.uk) – Although not exclusively a physics institution, it has a strong history of physics
- SPS National (http://www.spsnational.org) – Website of the Society of Physics Students

rue:Фізіка

Groundwater

Shipot, a common source of drinking water in a Ukrainian village.

Groundwater is water located beneath the ground surface in soil pore spaces and in the fractures of rock formations. A unit of rock or an unconsolidated deposit is called an aquifer when it can yield a usable quantity of water. The depth at which soil pore spaces or fractures and voids in rock become completely saturated with water is called the water table. Groundwater is recharged from, and eventually flows to, the surface naturally; natural discharge often occurs at springs and seeps, and can form oases or wetlands. Groundwater is also often withdrawn for agricultural, municipal and industrial use by constructing and operating extraction wells. The study of the distribution and movement of groundwater is hydrogeology, also called groundwater hydrology.

Typically, groundwater is thought of as liquid water flowing through shallow aquifers, but technically it can also include soil moisture, permafrost (frozen soil), immobile water in very low permeability bedrock, and deep geothermal or oil formation water. Groundwater is hypothesized to provide lubrication that can possibly influence the movement of faults. It is likely that much of the Earth's subsurface contains some water, which may be mixed with other fluids in some instances. Groundwater may not be confined only to the Earth. The formation of some of the landforms observed on Mars may have been influenced by groundwater. There is also evidence that liquid water may also exist in the subsurface of Jupiter's moon Europa.

The entire surface water flow of the Alapaha River near Jennings, Florida going into a sinkhole leading to the Floridan Aquifer groundwater

Aquifers

An *aquifer* is a layer of porous substrate that contains and transmits groundwater. When water can flow directly between the surface and the saturated zone of an aquifer, the aquifer is unconfined. The deeper parts of unconfined aquifers are usually more saturated since gravity causes water to flow downward.

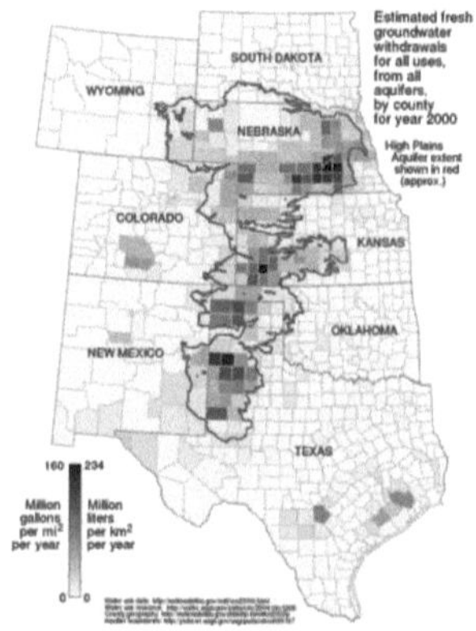

Groundwater withdrawal rates from the Ogallala Aquifer in the central U.S.

The upper level of this saturated layer of an unconfined aquifer is called the *water table* or *phreatic surface*. Below the water table, where generally all pore spaces are saturated with water is the phreatic zone.

Substrate with low porosity that permits limited transmission of groundwater is known as an *aquitard*. An *aquiclude* is a substrate with porosity that is so low it is virtually impermeable to groundwater.

A *confined aquifer* is an aquifer that is overlain by a relatively impermeable layer of rock or substrate such as an aquiclude or aquitard. If a confined aquifer follows a downward grade from its *recharge zone*, groundwater can become pressurized as it flows. This can create artesian wells that flow freely without the need of a pump and rise to a higher elevation than the static water table at the above, unconfined, aquifer.

The characteristics of aquifers vary with the geology and structure of the substrate and topography in which they occur. Generally, the more productive aquifers occur in sedimentary geologic formations. By comparison, weathered and fractured crystalline rocks yield smaller quantities of groundwater in many environments. Unconsolidated to poorly cemented alluvial materials that have accumulated as valley-filling sediments in major river valleys and geologically subsiding structural basins are included among the most productive sources of groundwater.

The high specific heat capacity of water and the insulating effect of soil and rock can mitigate the effects of climate and maintain groundwater at a relatively steady temperature. In some places where groundwater temperatures are maintained by this effect at about 50°F/10°C, groundwater can be used for controlling the temperature inside structures at the surface. For example, during hot weather relatively cool groundwater can be pumped through radiators in a home and then returned to the ground in another well. During cold seasons, because it is relatively warm, the water can be used in the same way as a source of heat for heat pumps that is much more efficient than using air.

Water cycle

Groundwater makes up about twenty percent of the world's fresh water supply, which is about 0.61% of the entire world's water, including oceans and permanent ice. Global groundwater storage is roughly equal to the total amount of freshwater stored in the snow and ice pack, including the north and south poles. This makes it an important resource which can act as a natural storage that can buffer against shortages of surface water, as in during times of drought.[1]

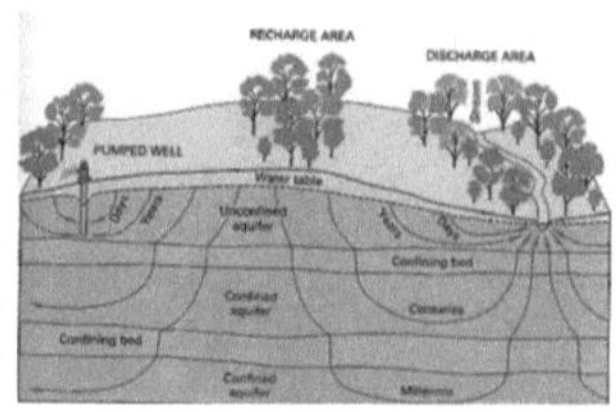

Relative groundwater travel times.

Groundwater is naturally replenished by surface water from precipitation, streams, and rivers when this recharge reaches the water table.

Groundwater can be a long-term 'reservoir' of the natural water cycle (with residence times from days to millennia), as opposed to short-term water reservoirs like the atmosphere and fresh surface water (which have residence times from minutes to years). The figure shows how deep groundwater (which is quite distant from the surface recharge) can take a very long time to complete its natural cycle.

The Great Artesian Basin in central and eastern Australia is one of the largest confined aquifer systems in the world, extending for almost 2 million km^2. By analysing the trace elements in water sourced from deep underground, hydrogeologists have been able to determine that water extracted from these aquifers can be more than 1 million years old.

By comparing the age of groundwater obtained from different parts of the Great Artesian Basin, hydrogeologists have found it increases in age across the basin. Where water recharges the aquifers along the Eastern Divide, ages are young. As groundwater flows westward across the continent, it increases in age, with the oldest groundwater occurring in the western parts. This means that in order to have travelled almost 1000 km from the source of recharge in 1 million years, the groundwater flowing through the Great Artesian Basin travels at an average rate of about 1 metre per year.

Issues

Overview

Certain problems have beset the use of groundwater around the world. Just as river waters have been over-used and polluted in many parts of the world, so too have aquifers. The big difference is that aquifers are out of sight. The other major problem is that water management agencies, when calculating the 'sustainable yield' of aquifer and river water, have often counted the same water twice, once in the aquifer, and once in its connected river. This problem, although understood for centuries, has persisted, partly through inertia within government agencies. In Australia, for example, prior to the statutory reforms initiated by the Council of Australian Governments water reform framework in the 1990s, many Australian States managed groundwater and surface water through separate government agencies, an approach beset by rivalry and poor communication.

The time lags inherent in the dynamic response of groundwater to development have generally been ignored by water management agencies, decades after scientific understanding of the issue was consolidated. In brief, the effects of groundwater overdraft (although undeniably real) may take decades or centuries to manifest themselves. In a classic study in 1982, Bredehoeft and colleagues[2] modelled a situation where groundwater extraction in an intermontane basin withdrew the entire annual recharge, leaving 'nothing' for the natural groundwater-dependent vegetation community. Even when the borefield was situated close to the vegetation, 30% of the original vegetation demand could still be met by the lag inherent in the system after 100 years. By year 500 this had reduced to 0%, signalling complete death of the groundwater-dependent vegetation. The science has been available to make these calculations for decades; however water management agencies have generally ignored effects which will appear outside the rough timeframe of political elections (3 to 5 years). Marios Sophocleous[2] argued strongly that management agencies must define and use appropriate timeframes in groundwater planning. This will mean calculating groundwater withdrawal permits based on predicted effects decades, sometimes centuries in the future.

As water moves through the landscape it collects soluble salts, mainly sodium chloride. Where such water enters the atmosphere through evapotranspiration, these salts are left behind. In irrigation districts, poor drainage of soils and surface aquifers can result in water tables coming to the surface in low-lying areas. Major land degradation problems of soil salinity and waterlogging result,[3] combined with increasing levels of salt in surface waters. As a consequence, major damage has occurred to local economies and environments.[4]

Four important effects are worthy of brief mention. First, flood mitigation schemes, intended to protect infrastructure built on floodplains, have had the unintended consequence of reducing aquifer recharge associated with natural flooding. Second, prolonged depletion of groundwater in extensive aquifers can result in land subsidence, with associated infrastructure damage – as well as (thirdly) saline intrusion.[5] Fourth, draining acid sulphate soils, often found in low-lying coastal plains, can result in acidification and pollution of formerly freshwater and estuarine streams.[6]

Another cause for concern is that groundwater drawdown from over-allocated aquifers has the potential to cause severe damage to both terrestrial and aquatic ecosystems – in some cases very conspicuously but in others quite imperceptibly because of the extended period over which the damage occurs.[7]

Overdraft

Groundwater is a highly useful and often abundant resource. However, over-use, or overdraft, can cause major problems to human users and to the environment. The most evident problem (as far as human groundwater use is concerned) is a lowering of the water table beyond the reach of existing wells. Wells must consequently be deepened to reach the groundwater; in some places (e.g., California, Texas and India) the water table has dropped hundreds of feet because of extensive well pumping. In the Punjab region of India, for example, groundwater levels have dropped 10 meters since 1979, and the rate of depletion is accelerating.[8] A lowered water table may, in turn, cause other problems such as groundwater-related subsidence and saltwater intrusion.

Wetlands contrast the arid landscape around Middle Spring, Fish Springs National Wildlife Refuge, Utah.

Groundwater is also ecologically important. The importance of groundwater to ecosystems is often overlooked, even by freshwater biologists and ecologists. Groundwaters sustain rivers, wetlands and lakes, as well as subterranean ecosystems within karst or alluvial aquifers.

Not all ecosystems need groundwater, of course. Some terrestrial ecosystems – for example, those of the open deserts and similar arid environments – exist on irregular rainfall and the moisture it delivers to the soil, supplemented by moisture in the air. While there are other terrestrial ecosystems in more hospitable environments where groundwater plays no central role, groundwater is in fact fundamental to many of the world's major ecosystems. Water flows between groundwaters and surface waters. Most rivers, lakes and wetlands are fed by, and (at other places or times) feed groundwater, to varying degrees. Groundwater feeds soil moisture through percolation, and many terrestrial vegetation communities depend directly on either groundwater or the percolated soil moisture above the aquifer for at least part of each year. Hyporheic zones (the mixing zone of streamwater and groundwater) and riparian zones are examples of ecotones largely or totally dependent on groundwater.

Aquifer drawdown or overdrafting and the pumping of fossil water increases the total amount of water within the hydrosphere subject to transpiration and evaporation processes, thereby causing accretion in water vapour and cloud cover, the primary absorbers of infrared radiation in the Earth's atmosphere. Adding water to the system has a forcing effect on the whole earth system, an accurate estimate of which hydrogeological fact is yet to be quantified.

A solution to over-use of groundwater

Irrigated agriculture is a big business in Asia. It accounts for 70 per cent of the world's irrigated land and 73 per cent of water used each year by the farming industry. As ageing large-scale surface irrigation schemes have become increasingly inefficient, and farmers have begun growing a wider range of crops requiring water on demand, the number of groundwater wells in India has exploded.[9] In 1960, there were fewer than 100,000 such wells; by 2006 the figure had risen to nearly 12 million.[10] In India, a possible solution to over-use of groundwater is emerging, known as 'groundwater recharge'. It involves capturing rainwater that would otherwise run-off, and using it to refill aquifers. Since 2000, the International Water Management Institute (IWMI) has been working with the Indian authorities to help improve the availability of water for agriculture in India. In 2006, India's finance minister invited

IWMI to submit policy recommendations based on its research on groundwater depletion. One of the key recommendations was to instigate a programme of recharging groundwater across the 65 per cent of India that has hard-rock aquifers. As a result, the Indian government allocated Rs 1800 crore (US$400million) to fund dug-well recharge projects (a dug-well is a wide, shallow well, often lined with concrete) in 100 districts within seven states where water stored in hard-rock aquifers has been over-exploited. These geological formations have a much lower capacity to store rainwater than alluvial areas with porous sand or clay rocks, hence being given priority. The money is sufficient to fund seven million structures to be installed on dug-wells to divert monsoon runoff. The structures include a de-siltation chamber, plus pipes to collect surplus rainwater and divert de-silted water from the chamber to the well.[11] As of the end of November 2009, funds amounting to Rs. 216.98 crore (including Rs. 199.98 crore as subsidy to beneficiaries and 17 crore for IEC/Capacity Building activities) had been released to the concerned states. Subsidies had been released to 566,637 beneficiaries.[12]

Subsidence

Subsidence occurs when too much water is pumped out from underground, deflating the space below the above-surface, and thus causing the ground to actually collapse. The result can look like craters on plots of land. This occurs because in its natural equilibrium state, the hydraulic pressure of groundwater in the pore spaces of the aquifer and the aquitard supports some of the weight of the overlying sediments. When groundwater is removed from aquifers by excessive pumping, pore pressures in the aquifer drop and compression of the aquifer may occur. This compression may be partially recoverable if pressures rebound, but much of it is not. When the aquifer gets compressed it may cause land subsidence, a drop in the ground surface. The city of New Orleans, Louisiana, is actually below sea level today, and its subsidence is partly caused by removal of groundwater from the various aquifer/aquitard systems beneath it. In the first half of the 20th century, the city of San Jose, California, dropped 13 feet from land subsidence caused by overpumping; this subsidence has been halted with improved groundwater management.

Seawater intrusion

Generally, in very humid or undeveloped regions, the shape of the water table mimics the slope of the surface. The recharge zone of an aquifer near the seacoast is likely to be inland, often at considerable distance. In these coastal areas, a lowered water table may induce sea water to reverse the flow toward the land. Sea water moving inland is called a saltwater intrusion. Alternatively, salt from mineral beds may leach into the groundwater of its own accord.

Mining

Sometimes the water movement from the recharge zone to the place where it is withdrawn may take centuries (see figure above). When the usage of water is greater than the recharge, it is referred to as *mining* water (the water is often called fossil water because of its geologic age). Under those circumstances it is not a renewable resource.

Pollution

Iron oxide staining caused by reticulation from an unconfined aquifer in karst topography. Perth, Western Australia.

Water pollution of groundwater, from pollutants released to the ground that can work their way down into groundwater, can create a contaminant plume within an aquifer. Movement of water and dispersion within the aquifer spreads the pollutant over a wider area, its advancing boundary often called a plume edge, which can then intersect with groundwater wells or daylight into surface water such as seeps and springs, making the water supplies unsafe for humans and wildlife. The interaction of groundwater contamination with surface waters is analyzed by use of hydrology transport models.

The stratigraphy of the area plays an important role in the transport of these pollutants. An area can have layers of sandy soil, fractured bedrock, clay, or hardpan. Areas of karst topography on limestone bedrock are sometimes vulnerable to surface pollution from groundwater. Earthquake faults can also be entry routes for downward contaminant entry. Water table conditions are of great importance for drinking water supplies, agricultural irrigation, waste disposal (including nuclear waste), wildlife habitat, and other ecological issues.

In the US, upon commercial real estate property transactions both groundwater and soil are the subjects of scrutiny, with a Phase I Environmental Site Assessment normally being prepared to investigate and disclose potential pollution issues. In the San Fernando Valley of California, Real estate contracts for property transfer below the Santa Susana Field Laboratory (SSFL) and eastward have clauses releasing the seller from liability for groundwater contamination consequences from existing or future water pollution of the Valley Aquifer.

Love Canal was one of the most widely known examples of groundwater pollution. In 1978, residents of the Love Canal neighbourhood in upstate New York noticed high rates of cancer and an alarming number of birth defects. This was eventually traced to organic solvents and dioxins from an industrial landfill that the neighbourhood had been built over and around, which had then infiltrated into the water supply and evaporated in basements to further contaminate the air. Eight hundred families were reimbursed for their homes and moved, after extensive legal battles and media coverage.

Another example of widespread groundwater pollution is in the Ganges Plain of northern India and Bangladesh where severe contamination of groundwater by naturally occurring arsenic affects 25% of water wells in the shallower of two regional aquifers. The pollution occurs because aquifer sediments contain organic matter (dead plant material) that generates anaerobic (an environment without oxygen) conditions in the aquifer. These conditions result in the microbial dissolution of iron oxides in the sediment and thus the release of the arsenic, normally strongly bound to iron oxides, into the water. As a consequence, arsenic-rich groundwater is often iron-rich, although secondary processes often obscure the association of dissolved arsenic and dissolved iron.

Ground Water Rule

In November 2006, the Environmental Protection Agency published the Ground Water Rule in the United States Federal Register. The EPA was worried that the ground water system would be vulnerable to contamination from fecal matter. The point of the rule was to keep microbial pathogens out of public water sources [13] The 2006 Ground Water Rule was an amendment of the 1996 Safe Drinking Water Act.

See also

- Groundwater models
- Groundwater flow
- Baseflow
- Seep (hydrology)
- Spring (hydrosphere)
- Water well
- Optimum water content for tillage

References

[1] "Learn More: Groundwater" (http://water.columbia.edu/?id=learn_more&navid=groundwater/). Columbia Water Center. . Retrieved 2009-09-15.

[2] Sophocleous, Marios (2002). "Interactions between groundwater and surface water: the state of the science". *Hydrogeology Journal* **10**: 52–67. doi:10.1007/s10040-001-0170-8.

[3] "Free articles and software on drainage of waterlogged land and soil salinity control" (http://www.waterlog.info). . Retrieved 2010-07-28.

[4] Ludwig, D.; Hilborn, R.; Walters, C. (1993). "Uncertainty, Resource Exploitation, and Conservation: Lessons from History" (http://landscape.forest.wisc.edu/courses/Landscape565spr01/Ludwig_etal1993.pdf). *Science* **260** (5104): 17–36. doi:10.1126/science.260.5104.17. JSTOR 1942074. PMID 17793516. .

[5] Zektser et al.

[6] Sommer, Bea; Horwitz, Pierre; Sommer, Bea; Horwitz, Pierre (2001). "Water quality and macroinvertebrate response to acidification following intensified summer droughts in a Western Australian wetland". *Marine and Freshwater Research* **52** (7): 1015. doi:10.1071/MF00021.

[7] Zektser, S.; Lo�Iciga, H. A.; Wolf, J. T. (2004). "Environmental impacts of groundwater overdraft: selected case studies in the southwestern United States". *Environmental Geology* **47** (3): 396–404. doi:10.1007/s00254-004-1164-3.

[8] Upmanu Lall. "Punjab: A tale of prosperity and decline" (http://blogs.ei.columbia.edu/water/2009/07/28/punjab-a-tale-of-prosperity-and-decline/). Columbia Water Center. . Retrieved 2009-09-11.

[9] Mukherji, A. Revitalising Asia's Irrigation: To sustainably meet tomorrow's food needs (http://www.iwmi.cgiar.org/Publications/Other/index.aspx) 2009, IWMI and FAO

[10] Chapter 10: The Groundwater Recharge Movement in India (http://www.iwmi.cgiar.org/publications/CABI_Publications/CA_CABI_Series/Ground_Water/protected/Giordano_1845931726-Chapter10.pdf), By Sakthivadivel, R. in *The Agricultural Groundwater Revolution*, Ed. Giordano, M. and Villholth, K, CABI Publications, 2007

[11] Influencing irrigation policy in India (http://www.iwmi.cgiar.org/Publications/Success_Stories/index.aspx), Success Story 6, 2010, International Water Management Institute, Colombo, Sri Lanka

[12] Initiatives in Water Resources (http://indiacurrentaffairs.org/initiatives-in-water-resources), Indian Current Affairs, December 30 2009, page accessed 5 May 2011

[13] Ground Water Rule (GWR) | Ground Water Rule | US EPA (http://water.epa.gov/lawsregs/rulesregs/sdwa/gwr/index.cfm). Water.epa.gov. Retrieved on 2011-06-09.

External links

- USGS Office of Groundwater (http://water.usgs.gov/ogw/)
- UK Groundwater Forum (http://www.groundwateruk.org)
- Status of Ground Water in India (http://www.indiawaterportal.org/channels/groundwater)
- The Groundwater Foundation (http://www.groundwater.org)
- Groundwater Information from the Coastal Ocean Institute (http://www.whoi.edu/page.do?pid=11929)
- US National Ground Water Association (http://www.ngwa.org)
- Connected Waters Initiative, University of New South Wales (http://www.connectedwaters.unsw.edu.au)
- American Water Resources Association (http://www.awra.org)
- Multimedia course on groundwater (http://www.indiawaterportal.org/multimedia.html)
- Bibliography on Water Resources and International Law (http://www.ppl.nl/index.php?option=com_wrapper&view=wrapper&Itemid=82) Peace Palace Library
- IGRAC, International Groundwater Resources Assessment Centre (http://www.igrac.net)
- Groundwater (http://www.hydrology.nl/key-publications/231-groundwater-geology-of-the-netherlands.html) Chapter on groundwater in the Netherlands, from 'Geology of the Netherlands'
- Arie S.Issar. The Evolution of Groundwater Exploitation Methods in the Middle East through History. (http://www.chaire-rome.hst.ulaval.ca/Docs_revue/html/revue_etat_questions_arie_issar_evolution_groundwater_exploitation_methods.htm)
- US Army Geospatial Center (http://www.agc.army.mil/) — For information on OCONUS surface water and groundwater.

Petrography

Petrography is a branch of petrology that focuses on detailed descriptions of rocks. Someone who studies petrography is called a **petrographer**. The mineral content and the textural relationships within the rock are described in detail. Petrographic descriptions start with the field notes at the outcrop and include megascopic description of hand specimens. However, the most important tool for the petrographer is the petrographic microscope. The detailed analysis of minerals by optical mineralogy in thin section and the micro-texture and structure are critical to understanding the origin of the rock. Electron microprobe analysis of individual grains as well as whole rock chemical analysis by atomic absorption or X-ray fluorescence are used in a modern petrographic lab. Individual mineral grains from a rock sample may also be analyzed by X-ray diffraction when optical means are insufficient. Analysis of microscopic fluid inclusions within mineral grains with a heating stage on a petrographic microscope provides clues to the temperature and pressure conditions existent during the mineral formation.

Methods of investigation

Macroscopic characters

The macroscopic characters of rocks, those visible in hand-specimens without the aid of the microscope, are very varied and difficult to describe accurately and fully. The geologist in the field depends principally on them and on a few rough chemical and physical tests; and to the practical engineer, architect and quarry-master they are all-important. Although frequently insufficient in themselves to determine the true nature of a rock, they usually serve for a preliminary classification, and often give all the information needed.

With a small bottle of acid to test for carbonate of lime, a knife to ascertain the hardness of rocks and minerals, and a pocket lens to magnify their structure, the field geologist is rarely at a loss to what group a rock belongs. The fine grained species are often indeterminable in this way, and the minute mineral components of all rocks can usually be ascertained only by microscopic examination. But it is easy to see that a sandstone or grit consists of more or less rounded, water-worn sand grains and if it contains dull, weathered particles of feldspar, shining scales of mica or small crystals of calcite these also rarely escape observation. Shales and clay rocks generally are soft, fine grained, often laminated and not infrequently contain minute organisms or fragments of plants. Limestones are easily marked with a knife-blade, effervesce readily with weak cold acid and often contain entire or broken shells or other fossils. The crystalline nature of a granite or basalt is obvious at a glance, and while the former contains white or pink feldspar, clear vitreous quartz and glancing flakes of mica, the other shows yellow-green olivine, black augite, and gray stratiated plagioclase.

Other simple tools include the blowpipe (to test the fusibility of detached crystals), the goniometer, the magnet, the magnifying glass and the specific gravity balance.[1]

Microscopic characteristics

When dealing with unfamiliar types or with rocks so fine grained that their component minerals cannot be determined with the aid of a hand lens, a microscope is used. Characteristics observed under the microscope include colour, colour variation under plane polarised light (pleochroism, produced by the lower Nicol prism, or more recently polarising films), fracture characteristics of the grains, refractive index (in comparison to the mounting adhesive, typically Canada Balsam), and optical symmetry (birefringent or isotropic). *In toto*, these characteristics are sufficient to identify the mineral, and often to quite tightly estimate its major element composition. The process of identifying minerals under the microscope is fairly subtle, but also mechanistic - it would be possible to develop an identification key that would allow a computer to do it. The more difficult and skilful part of optical petrography is identifying the interrelationships between grains and relating them to features seen in hand specimen, at outcrop, or in mapping.

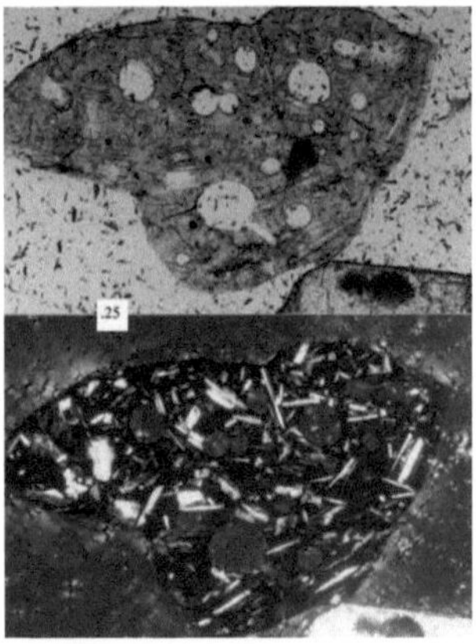

Photomicrograph of a volcanic sand grain; upper picture is plane-polarized light, bottom picture is cross-polarized light, scale box at left-center is 0.25 millimeter.

Separation of components

Separation of the ingredients of a crushed rock powder to obtain pure samples for analysis is a common approach. It may be performed with a powerful, adjustable-strength electromagnet. A weak magnetic field attracts magnetite, then haematite and other iron ores. Silicates that contain iron follow in definite order—biotite, enstatite, augite, hornblende, garnet, and similar ferro-magnesian minerals are successively abstracted. Finally, only the colorless, non-magnetic compounds, such as muscovite, calcite, quartz, and feldspar remain. Chemical methods also are useful.

A weak acid dissolves calcite from crushed limestone, leaving only dolomite, silicates, or quartz. Hydrofluoric acid attacks feldspar before quartz and, if used cautiously, dissolves these and any glassy material in a rock powder before it dissolves augite or hypersthene.

Methods of separation by specific gravity have a still wider application. The simplest of these is levigation—treatment by a current of water. Levigationis extensively employed in mechanical analysis of soils and treatment of ores, but is not so successful with rocks, as their components do not, as a rule, differ greatly in specific gravity. Fluids are used that do not attack most rock-forming minerals, but have a high specific gravity. Solutions of potassium mercuric iodide (sp. gr. 3.196), cadmium borotungstate (sp. gr. 3.30), methylene iodide (sp. gr. 3.32), bromoform (sp. gr. 2.86), or acetylene bromide (sp. gr. 3.00) are the principal fluids employed. They may be diluted (with water, benzene, etc.) or concentrated by evaporation.

If the rock is granite consisting of biotite (sp. gr. 3.1), muscovite (sp. gr. 2.85), quartz (sp. gr. 2.65), oligoclase (sp. gr. 2.64), and orthoclase (sp. gr. 2.56), the crushed minerals float in methylene iodide. On gradual dilution with benzene they precipitate in the order above. Simple in theory, these methods are tedious in practice, especially as it is common for one rock-making mineral to enclose another. However, expert handling of fresh and suitable rocks yields excellent results.[1]

Chemical analysis

In addition to naked-eye and microscopic investigation, chemical research methods of are of great practical importance to the petrographer. Crushed and separated powders, obtained by the processes above, may be analyzed to determine chemical composition of minerals in the rock qualitatively or quantitatively. Chemical testing, and microscopic examination of minute grains is an elegant and valuable means of discriminating between mineral components of fine-grained rocks.

Thus, the presence of apatite in rock-sections is established by covering a bare rock-section with ammonium molybdate solution. A turbid yellow precipitate forms over the crystals of the mineral in question (indicating the presence of phosphates). Many silicates are insoluble in acids and cannot be tested in this way, but others are partly dissolved, leaving a film of gelatinous silica that can be stained with coloring matters, such as the aniline dyes (nepheline, analcite, zeolites, etc.).

Complete chemical analysis of rocks are also widely used and important, especially in describing new species. Rock analysis has of late years (largely under the influence of the chemical laboratory of the United States Geological Survey) reached a high pitch of refinement and complexity. As many as twenty or twenty-five components may be determined, but for practical purposes a knowledge of the relative proportions of silica, alumina, ferrous and ferric oxides, magnesia, lime, potash, soda and water carry us a long way in determining a rock's position in the conventional classifications.

A chemical analysis is usually sufficient to indicate whether a rock is igneous or sedimentary, and in either case to accurately show what subdivision of these classes it belongs to. In the case of metamorphic rocks it often establishes whether the original mass was a sediment or of volcanic origin.[1]

Specific gravity

Specific gravity of rocks is determined by means of a balance and pycnometer. It is greatest in rocks with the most magnesia, iron, and heavy metals—least in rocks rich in alkalis, silica, and water. It diminishes with weathering and, generally, highly crystalline rocks have higher specific gravity than wholly or partly vitreous rocks of the same chemical composition. The specific gravity of the commoner rocks ranges from about 2.5 to 3.2.[1]

Archaeological applications

Archaeologists use petrography to identify mineral components in pottery. This information ties the artifacts to geological source areas for the clay used and the rock fragments (usually called "temper" or "aplastics") potters often add to modify the clay's properties. This provides insight into how potters selected and used local and non-local resources, and helps archaeologists determine whether pottery found in a particular location was locally produced or traded from elsewhere. In turn, this kind of information (in combination with other evidence) can support inferences about settlement patterns, group and individual mobility, and social contacts or trade networks. In addition, an understanding of how certain minerals are altered at specific temperatures can allow archaeological petrographers to infer aspects of the ceramic production process itself, such as minimum and maximum temperatures reached in the original firing of the pot.

See also

- Ceramic petrography

References

[1] This article incorporates text from a publication now in the public domain: Chisholm, Hugh, ed (1911). "Petrology". *Encyclopædia Britannica* (11th ed.). Cambridge University Press.

External links

- Atlas of Rocks, Minerals, and Textures (http://geo.web.ru/IgMetAtlas/) Petrographical description of rocks and minerals

Economic geology

Economic geology is concerned with earth materials that can be used for economic and/or industrial purposes. These materials include precious and base metals, nonmetallic minerals, construction-grade stone, petroleum minerals, coal, and water. The term commonly refers to metallic mineral deposits and mineral resources. The techniques employed by other earth science disciplines (such as geochemistry, mineralogy, geophysics, and structural geology) might all be used to understand, describe, and exploit an ore deposit.

Economic geology is studied and practiced by geologists. However it is of prime interest to investment bankers, stock analysts and other professions such as engineers, environmental scientists, and conservationists because of the far-reaching impact that extractive industries have on society, the economy, and the environment.

Mineral resources

Mineral resources are concentrations of minerals significant for current and future societal needs. Ore is classified as mineralization economically and technically feasible for extraction. Not all mineralization meets these criteria for various reasons. The specific categories of mineralization in an economic sense are:

- *Mineral occurrences* or prospects of geological interest but not necessarily economic interest
- *Mineral resources* include those potentially economically and technically feasible and those that are not
- *Ore reserves*, which must be economically and technically feasible to extract

Ore geology

Geologists are involved in the study of ore deposits, which includes the study of ore genesis and the processes within the Earth's crust that form and concentrate ore minerals into economically viable quantities.

Study of metallic ore deposits involves the use of structural geology, geochemistry, the study of metamorphism and its processes, as well as understanding metasomatism and other processes related to ore genesis.

Ore deposits are delineated by mineral exploration, which uses geochemical prospecting, drilling and resource estimation via geostatistics to quantify economic ore bodies. The ultimate aim of this process is mining.

Coal and petroleum geology

See main articles Coal and Petroleum geology

The study of sedimentology is of prime importance to the delineation of economic reserves of petroleum and coal energy resources.

References

- U.S. Geological Survey Circular 831, Principles of a Resource/Reserve Classification for Minerals (PDF format) [1]

See also

- Important publications in economic geology
- Mineral resource classification
- Ore
- Ore genesis
- Coal

References

[1] http://www.ut.blm.gov/NewsReleases/nrjan16USGSCircular831.pdf

Article Sources and Contributors

Heinz Ebert *Source*: http://en.wikipedia.org/w/index.php?title=Heinz_Ebert *Contributors*: 1ForTheMoney, Crusio, Geologist, Janggeom, Lockley, 4 anonymous edits

Natural history *Source*: http://en.wikipedia.org/w/index.php?title=Natural_history *Contributors*: 100110100, A8UDI, Abdullais4u, Adamsan, Alan Liefting, Andrewrp, Andy Marchbanks, Angr, Atulsnischal, Awickert, BanyanTree, Brian Crawford, Brian0918, Brunnock, CJLL Wright, Cablanchet, ChrisGriswold, Clementsp, Clicketyclack, Colincbn, Conscious, Curtis Clark, Cyprus2k1, DV8 2XL, DaGizza, David Schaich, DirkvdM, Dougweller, Dr.Bastedo, Dudester11, EagleFan, Earthdirt, Epbr123, Erianna, Evolauxia, Fabartus, Foxleas, Francis Schonken, Franzeska, Gaurav, Gdr, GeorgeLouis, Geothermal, Gfoley4, Giraffedata, Greyhood, Gtrmp, Haymaker, Headbomb, Hike395, Hmains, Icairns, Invertzoo, J.delanoy, Jacoahbrothers, Jacobko, Jagged 85, Jlittlet, Jmabel, Jni, JonRichfield, Jondel, Joseph Solis in Australia, Jrockley, KPalicz, Kiesewetter, Kitegirl15, Koavf, Kozuch, LeaveSleaves, Leonard G., Leszek Jańczuk, Lexor, Lightmouse, Lorenzarius, LouI, Lquilter, Luk, Lukobe, Lynxmb, M1ss1ontomars2k4, Macdonald-ross, Mahanga, Malcolm Farmer, Maxis ftw, Mesoderm, Michaelritchie200, Mikenorton, Minimax, Mzalamea, Nate Silva, Nemetona, Nonenmac, Ojigiri, Oxymoron83, Parkranger22, Pertusaria, Peter coxhead, Phe, Phlebas, Physis, Pietro6, Pigsonthewing, RJBurkhart, Ragesoss, Rdsmith4, RedWolf, Reddi, RenniePet, Richard001, Richdeline, Ripepette, Rjwilmsi, Rosi Crane, Rtully1, Salanth, Samsara, SchoonerGuy, Scilit, Sciurinæ, SergeyJ, Shyamal, Steven J. Anderson, Sven Manguard, Technopat, Tewksjj, The way, the truth, and the light, Tmizuk, Tobby72, Tsiaojian lee, Twas Now, Unscented, Utcursch, Valérie75, VanishedUser314159, Viriditas, VivaEmilyDavies, Wapiti889, Waterguy, Wavelength, Wetman, WolfmanSF, Woohookitty, ZhaneMcBrown, 124 anonymous edits

Geologist *Source*: http://en.wikipedia.org/w/index.php?title=Geologist *Contributors*: 65.68.87.xxx, 9693, AaronDavid, Ahasuerus, Aitias, Ajl772, Altenmann, Andre Engels, Anetode, Antandrus, Awickert, Ayla, Bart133, Bassbonerocks, BigCow, Bobo192, Borgx, Calor, Can't sleep, clown will eat me, Canadian Paul, Caspertheghost, CattleGirl, Cometstyles, Conversion script, Courcelles, D, DARTH SIDIOUS 2, DVD R W, Darishante, Darth Panda, DavidLevinson, Dentren, DerHexer, Discospinster, DocWatson42, Drillerguy, Drsarno, Eastlaw, Ekabhishek, Epbr123, Etacar11, Evans1982, Excirial, FisherQueen, Frangibility, GeoPotinga, Geohumphrey, Geologyguy, Geoz, Gogo Dodo, Goldfritha, GooPC, GregorB, Greyhood, Grillo, Gscshoyru, Gwernol, Hawkbreeze, Icestorm815, Iridescent, IrisKawling, IronGargoyle, It Is Me Here, Ixfd64, J.delanoy, Jammat, Jimcgeo, Jstuby, Juliancolton, Kcowolf, Kingpin13, Kobinks, Korg, Ktsquare, Levineps, Linnell, MER-C, MPF, Magister Mathematicae, Marek69, Masterusagi, Maths222, Matthew Yeager, Mav, Maxis ftw, Meekywiki, Mikenorton, Minimistro, Monterey Bay, Mutinus, NawlinWiki, NeoJustin, Nihonjoe, Numbsilence, OllieFury, PGHGEOLOGIST, Palica, PaulGarner, Phaldo, Piano non troppo, Pinethicket, Pkrembs, Poopoopoopoo167, Primate, Qfl247, Quentar, Qxz, R'n'B, RaCha'ar, Red Thrush, RedWolf, Res2216firestar, Rolinator, RoyBoy, Sasha Kopf, Shanel, Signalhead, Siim, SimonP, Slf8261, Smalljim, StuffOfInterest, Syndandrome, THEN WHO WAS PHONE?, Tellyaddict, The Thing That Should Not Be, Thingg, Thomas Larsen, Thue, Thunderthief, Tide rolls, Tillman, Tommy2010, Treisijs, Tridy, Uyanga, Vonones, Vrenator, Vter4life, Wahid alig, West.andrew.g, Westsider, Wikaholic, Wiki alf, WikiLaurent, Wikipe-tan, Willking1979, Wilson44691, Wimt, Ww2censor, Yamaguchi, Zamphuor, Zap Rowsdower, , 359 anonymous edits

Teacher *Source*: http://en.wikipedia.org/w/index.php?title=Teacher *Contributors*: 0dd1, 2D, 4-409r-0, 4twenty42o, 5 albert square, 53180, 7, 928s4, A Sengupta, A little insignificant, A3RO, ABF, AaronY, Abc518, Abdowiki, Abdullais4u, Abhiras, Acalamari, Acather96, Achmed1235, Adamfinmo, Adantonio, Addshore, AdelaMae, AdjustShift, Adrian, Ageekgal, AgentPeppermint, Ahoerstemeier, Ahuitzotl, Aidakin, Airplaneman, Aitias, Alansohn, Aleksd, Alex.muller, AlexiusHoratius, Alexwillerick, Alfrodull, Allibobally, AlotToLearn, Alphaboi867, Amber.bush, Amtoj amtoj, Anakin101, Andre Engels, Andrea105, Andres, Andres rojas22, Andrewa, Andrewpmk, Andrewthemaster22, Andycjp, Angie Y., Angrysockhop, Anna Lincoln, Annie 54, Anonymous Dissident, Antandrus, Antd129, Antdoran, AnthonyUK, Antonio Lopez, Anveo, Apostrophe, Arakunem, ArcAngel, Arctic Night, Ardara, ArglebargleIV, Arliedabb, Asirhan45, Askahrc, Asst2drsalary, Autocracy, Avenged Eightfold, Avicennasis, Avillia, Avoided, Azgorodin, B, BAKA4lyfe, BW95, Baa, Babyblushes, Babyblusshes, Bad edits r dumb, Badbilltucker, Banno, Barek, Baronnet, BarretBonden, Bart133, Bartledan, Bass fishing physicist, Bbriggs1, BeSherman, Beano, Beemer69, Beijingbear, Belinrahs, Bencherlite, Benzachphill, Bhavargarg, Bhawani Gautam, Bibliomaniac15, Big Bird, Big J Ross, BillFlis, Billyjr560, Bkonrad, BlackCat00, Blazingcheeser, Bluerasberry, Bob rulz, Bobblewik, Bobbyboberson, Bobo1235, Bobo192, Bobrocks222, Bodhisattvaspath, Bongwarrior, Boobala2000, Bouefboeufhupesisnub, Bremerenator, BrendelSignature, Brian Crawford, Brian the Editor, Brianga, Brideshead, BryanG, Bsadowski1, Buckleddown, CHJL, CIreland, COMPFUNK2, CaesarGJ, Calliopejen1, Caltas, CalumH93, Calvin 1998, Camembert, Camille1993, Can't sleep, clown will eat me, CanadianLinuxUser, Canjth, Capricorn42, Captain Virtue, Captain-n00dle, Captdoc, CardinalDan, Carloscarlos8675309, Carmine1235, Carnildo, Catgut, Cause of death, Cbradshaw, CenozoicEra, Certz, Chameleon, Chanlord, Chanting Fox, Charlesdrakew, Chase me ladies, I'm the Cavalry, Chasingsol, Chavash, Cheesefred02, Chenght, Chester Markel, ChicagoTeacher, Chicoisadog, Choralw, Choster, Chris Heiner, ChrisHodgesUK, Chrissycal, Chriswiki, Ciddler, Ckamaeleon, Ckatz, Clarince63, Clayton246, Click23, Closedmouth, Clovis Sangrail, Cmoneylong, Cobaltbluetony, Cold Water, Comet1, Cometstyles, CommonsDelinker, Connor75437, Cookie90, Coolchick19190, Corbit, Correogsk, Corvus cornix, Courcelles, Cp111, Cpl Syx, Csvinod, Cubs197, Curuinor, Cyktsui, D3ath2people, DANNYTHEKIDDY, DARTH SIDIOUS 2, DDerby, DJ Clayworth, DVD R W, Da1nonli, Dadude3320, Dakart, Dane Sorensen, Danger, DanielCD, Dapurpoopnub, Darth Panda, DarthShrine, DarthVader, Dave1185, David kim, DavidOaks, Davidkinnen, Dawn Bard, Dbiel, Dchall1, Deavenger, Dekimasu, Dekisugi, Deleter159, Delldot, Dentren, Deor, DerHexer, Derg4, Dermo69, Dethme0w, Deutschgirl, Dfrg.msc, Dialectric, Digitalme, Dinopup, Discospinster, DivineAlpha, Doc Tropics, Doctorateguy, Doctormanhattan, Docu, Doczilla, Dogposter, Dorvaq, DougsTech, Doulos Christos, Dreadstar, Drmies, Dsp13, Dudeedud, Duminda Dahanayake, Dweller, Dysepsion, Dysprosia, ESkog, EagleFan, EamonnPKeane, EchetusXe, Edbayley07, Edfen, Edgarde, Edward, Edward321, Eggplant123, Egmontaz, Eisnel, El C, Elec1000, Elinnea, Elipongo, Elvisfan2095, Epbr123, EphraimL, Erianna, Eric.dane, Erik9, Escape Orbit, Espresso Addict, Etacar11, Etan J. Tal, Etaoin, Ethan c.00, Everyguy, Everyking, Experiential, Extransit, FF2010, Fabulousfaison, Falcon8765, Fan-1967, Fantabulusfranci, Faradayplank, FastLizard4, Fasten, Faygo lover, Fearfulpenguin, Festus68, Finlay McWalter, Finoopy, Fishal, FisherQueen, Fishiehelper2, Fishonfire, Flewis, Flibirigit, Footwarrior, Frankenpuppy, Frecklefoot, Freechild, FreplySpang, Fuhghettaboutit, Fuzheado, GDallimore, Gabe64, Gail, Gaius Cornelius, Gareth E Kegg, Garfield226, Gears1235, Gene.arboit, Gfoley4, Ggfre, Ghooldy13, GiantPea1, Gil Gamesh, Gilliam, Gimboid13, Glane23, Gogo Dodo, Goh wz, GoldKanga, GraemeL, Greatal386, Greatgreenwhale, Greudin, Griffinofwales, Grim23, Gsp8181, Gtg204y, Guaka, Guy M, Guyzero, Gymfliperoo87, Gzkn, Hadal, Haham hanuka, Hal8999, Halospawn, Hamlet cat10, Hannefrogs, HarlandQPitt, Harryboyles, Hdt83, Heaven's Wrath, Hello32020, Henryodell, Herostratus, Hobartimus, Holdspa, Horacenew, Hordaland, HoundofBaskersville, Hpluver429, Hse45, Hshiek, Hughcharlesparker, Hussein962, Hut 8.5, Hutcher, Hydrogen Iodide, I already forgot, ITGSEETest, Iamg, IceMan4Life, Ichormosquito, Ifidieb4iwake, Igoldste, Ihatepeanuts67, Imafishgod, Imamaine, Imnewhere0, ImperatorExercitus, Imperial Monarch, Indon, Influence it, Infrogmation, Insanity Incarnate, Intelmachinehigh, Inter, Into The Fray, Iridescent, Iss246, It Is Me Here, Ivenius, Ixfd64, J. J. F. Nau, J.delanoy, JForget, JFreeman, JHeinonen, JMGPannal, Jack1993992, Jackronayne, Jacobisgay, JakkOne, JamieJones, Jaswal.raj, Javert, JavierMC, Jaymetzbaby, Jazzychick, Jblow21, Jbrennus, Jc37, Jeanenawhitney, Jeaniepang, Jebus989, Jeet0213, Jeff G., Jemijohn, Jessicavaz, Jevergreen, Jgoettisheim, Jillian9, Jimbood, Jimothym.r., Jimp, JivaGroup, Jivecat, Jjscone, Jklin, Jmeehan3, Joana92, JoanneB, Joaquindesalva, Joel7687, John Wilterman, JohnCD, Johnnyp77, Johnthebaker, JohnyDog, Jonel, Joriki, Joseph Solis in Australia, Jossi, Jpgordon, Juggleandhope, Juliancolton, Jusdafax, Jusjih, KCogburn, KF, KGasso, Kablammo, Kakadinho2210, Kanonkas, Kappus, Karenjc, Karimmariey, Kartano, Katalaveno, Katanada, Katieh5584, Kayau, Kcww, Keilana, KeithMatthew, Kencaesi, Kesac, Keunseokhur, Khayman, KibaKibbles, Killerkat90, Kilmer-san, Kingoomieiii, Kingpin13, Kintetsubuffalo, KnightLago, KnowledgeOfSelf, Koavf, Kreen, Krylonblue83, Kubigula, Kukini, Kungfuadam, Kuru, Kvdveer, Kwsn, LAX, LOL, La goutte de pluie, Lacort, Lacqui, Lalala3413, Latka, Leafyplant, LedgendGamer, Leifs1, Lessonlogs, Lewisskinner, LiamDXson123, LibLord, Lightdarkness, Lightmouse, Linaliang1401, Linzgirl85, LittleOldMe, Lkent 009, Lkent-06, LocoRoo, Longhair, Lord Bob, Lord Casus, Loren.wilton, Lsy098, LuigiManiac, Luk, LukeSurl, Luna Santin, Lurcher66, MER-C, MKoltnow, MONGO, MPorciusCato, Maddie!, Madhero88, Magioladitis, Magister Mathematicae, MagnaVox, Majorclanger, Majorly, Malcolma, Man vyi, Mandarax, Manthasam666, Marek69, Mark7-2, Marker540, Marshman, Martarius, Martin451, Martinp23, Masshaj, Master of Puppets, MasterOfTheXP, Math.angel, Matt Crypto, Matterfoot, Maurice Carbonaro, Maxim, Maximus Rex, Mazin07, Mboverload, McSly, Meekywiki, Megaboz, Melinawwe55, Menchi, Meneguce, Mentifisto, Metron4, Mgdowney, Mhking, Michael Hardy, Mike Rosoft, MindstormsKid, Miranda, Misterx2000, MithrandirAgain, Mjklin, Mmmkay35, Mmzi, Moonlight lonely, Moonriddengirl, Moralis, Moreschi, Morwen, Motley Crue Rocks, Moukity, Mr Adequate, Mr. Billion, Mr.K., MrJones, MrMurph101, MrOllie, Mrholybrain, Mwilso24, Mwuahahah, Myasuda, Mygerardromance, Mysid, N5iln, NHRHS2010, NSH001, NY-Bucks, Nachoman-au, Nadavspi, Naerii, Nakon, Nandesuka, Naq, NawlinWiki, Neelix, Nefariousski, Nemu, Nessalc, Netoholic, Neurolysis, Neverquick, NewEnglandYankee, Newnoise, Nezzadar, Nicholasnice, NickMartin, Nilja87, Nivix, Nleonardi, Noctibus, Noidontliekmudkipz, Nono64, NormalPersonEveryday, Norman1728, Notinasnaid, Notwgirl, Nsaa, Nubiatech, Nuttycoconut, O3scotto, Ocaasi, Ohiotap, Ohnoitsjamie, Olegwiki, Oli Filth, OllieFury, Omg441, Omonoid, One Way Motorway, Onorem, Optiquest222, Orangemike, Orangutan, Oscarthecat, Oskilian, Ott, Owen, OwenBlacker, OwenX, Oxnar9, Oxymoron83, Ozguy89, PCHS-NJROTC, PM100, Paddyboy315, Paganpan, Pahc, Pairadox, Pakaran, Palinb, Papa Smurf11, Papermate939, ParticleMan, Pastejohn, Paul August, Paulsprecker, Pavel Vozenilek, Pedro, Penbat, Pepper, Perfect Proposal, Pernogr, Persian Poet Gal, Peruvianllama, Peter Karlsen, Peterbrass, Peteweez, Pfctdayelise, Pgk, Phantomsteve, Pharaoh of the Wizards, Phenz, Philip Trueman, PhilipO, Philippe, Phoenix1304, Picnic29, Pigman, Pigsonthewing, Pinethicket, Pinkadelica, Pinkz 0x, Pjacobi, Pluke, Pmlineditor, Poindexter Propellerhead, PokeMaster45, Poop2556, Porqin, Porturology, Possum, Postdeleter, Ppntori, Prashanthns, Pseudomonas, Pstanton, Puchiko, Pundit, Pyrrhus16, Quadricode, Quantumor, Quercus basaseachicensis, Qxz, R'n'B, RA0808, Rachim, Radiant!, RadiantRay, RadioFan, Ralissa--, RandallFlagg Scotland, Random Passer-by, Ratout555, Ravenlord71, Razorflame, Rcawsey, Rculatta, Rdsmith4, Reach Out to the Truth, Reaper Eternal, Red, Reddi, Redskins09, Redwolfjoy, Reeveorama, ReinforcedReinforcements, Replacemats, Requestion, Res2216firestar, Retardedchickenwing, RexNL, Rgoodermote, Rhmorrow, Rich Farmbrough, Richard001, RichardF, Richhoncho, Rjd0060, Rjwilmsi, Robminchin, Rodhullandemu, Rolandg, Roleplayer, Rosser1954, Roving Wordslinger, Rrburke, Rtyq2, Rvde1996, RyanParis, SHTR, Saintcatherinelaguna, Saiyanlord, Samlyn.josfyn, Samwilson001, Sander123, Sasquatch, Sbuks17, Sceptre, SchfiftyThree, Schmiteye, SchnitzelMannGreek, School2190, Scifi30, Scottmsg, Sean D Martin, Sean.hou, Secretlondon, Sesmith, Sexyj Southards, Shadowjams, Shagrs11, Shalom Yechiel, Shanahan14, Shanes, Sharadaprasad, Sharkface217, Shnupbups, Sigurd Dragon Slayer, Simi2525, Sionus, Sjakkalle, Skaisgoodnotdead, Skarebo, Skimpy limp, SkyBoxx, Skyrocket, SleepyDan, SlimVirgin, Smadge1, Smalljim, Smithjd8, Smitsrr, SnappingTurtle, Snigbrook, Snooboysnoo, Snowolf, Soapandsoap, Solipsist, Someguy1221, Sonnysmith070795, Sophus Bie, Souljaboyent, Specialkay985, Specs112, Spiderman1235, Splash, Srose, Staffwaterboy, StanStandard, Starmoonman, Stephen Gilbert, Stephen Turner, Stephenb, Steve Smith, Steven Zhang, Stevenmitchell, Stevertigo, Stevietheman, Stickyboy1970, Storm Rider, Stwalkerster, Suffusion of Yellow, SuperHamster, SuperTycoon, Supperflash93, Sweeny111, Szyslak, THEN WHO WAS PHONE?, TITIEGIRL, TKFMPardus, Tafkam, TakuyaMurata, TamiquiaSimon, Tank 17, Tannkremen, Tapir Terrific, Taranet, Tbhotch, Tcncv, Teacher24, TeacherA, Teacherladyny, TeachersPet11, Teachr2001, TechBear, Terence, Terminator50, Terryn3, TexasAndroid, Thanscom84, Thatguy42, Thatguyflint, The Behnam, The Evil IP address, The Placebo Effect, The Thing That Should Not Be, The Transhumanist, The president of east africa, TheDJ, TheDestitutionOfOrganizedReligion, TheSuperBrain2, TheUserFormallyKnownAsEmkaer, Thedarkestclear, Themfromspace, Theoneintraining, Thimaeus, Thingg, Throwaway85, Thucydides of Thrace, Tide rolls, Tim10000, Tinton5, Tinyrock, Tmothyh, Tobby72, Tobias Conradi, Tomatoivy, Tommy2010, Tompw, Tomwsulcer, Toreau, Tostie14, Touch Of Light, Treisijs, Trendelltom, Triona, Trooper206, TrubZ, Tthaas, TuukkaH, Tvoz, Tvtrekker, Twlight532, Tyfster19, Tyler, Tyranno324, Uhhhhh, Uirauna, UkPaolo, Uncle Dick, Uncle Milty, Underdogggg, Urastink145, UserDoe, Utcursch, Uzecon,

VMS Mosaic, Vald, Valley2city, Vanderbilt8, Vanesaperu, Vespristiano, Viltharis, Violetriga, Voyagerfan5761, Vrenator, W Tanoto, WAvegetarian, WJetChao, WadeSimMiser, Washdivad, Wavelength, Wayne Slam, Wcbpolish, Weaponofmassinstruction, Wereon, Whicky1978, Who then was a gentleman?, WikiDan61, Wikieditor06, Wikigirlpga, Wikilibrarian, Wildt, Willandmartin, Willgrulich, Willking1979, Wimt, Wintonian, Wmahan, Wolfgang1018, Woohookitty, Wtmitchell, Ww2censor, Wæng, X3kkatee, Xact, Xezbeth, Xiaphias, XoPRCutie23Xo, Xp54321, Xteenb, Yamamoto Ichiro, Yermom123456, Yerpo, Yesukai, Yjones, Yohoroho, Yotimbob, YuTanaka, Yuckfoo, Yukikanou, ZachMartens, Zain Ebrahim111, Zalgo, Zanimum, Zauker, Zhou Yu, Zizonus, Znevetsch, Zzuuzz, Александър, ירין, 2024 anonymous edits

Chemistry *Source*: http://en.wikipedia.org/w/index.php?title=Chemistry *Contributors*: .mdk., 08walta, 194.200.130.xxx, 205.188.199.xxx, 64squares, 94chauj, A Doon, A Macedonian, AGK, APH, Acegikmo1, Acerperi, Adam Bishop, Adashiel, AdjustShift, Adminsr, Adqam, Adventurer, Aeluwas, Ahoerstemeier, Aiowe ;nhg8ohegpo8haeog, Ajrocke, Akanemoto, Alan Liefting, Alex S, Alhutch, Alison, Amcfreely, AmiDaniel, Amire80, Ams80, Anaraug, Ancheta Wis, Andrewpmk, Andris, Andy85719, Andycjp, Andyso, Angela, Anger22, Ann Stouter, Anonymous anonymous, Antandrus, Apha, ArglebargleIV, Arjun01, Arkuat, Arnon Chaffin, Art Carlson, Astrochemist, Astrovega, Atepnurdjaman, AuGold, AustinZ, AxelBoldt, Azareal7, BRG, Bact, Ballista, Ban Bridges, Barakta, Barticus88, Basharh, Basketballcooliog, Bduke, Beetstra, Ben D., Ben-Zin, Ben.c.roberts, Bender235, Benjah-bmm27, Bensaccount, Bfesser, Bidgee, Bill Nye the wheelin' guy, Bjf, Blackpower, Bluemask, Bmeguru, Bmicomp, Bob, Bob the ducq, Bobblewik, Bobby1011, Bobo192, Bongwarrior, Bonnerman, BookSquirrel, Boothy443, BorgQueen, Borislav, Bornhj, Braddodson, Bradeos Graphon, Braindrain0000, Brandon5485, Brent45801, Brian Pearson, Brian0918, Brianga, Brion VIBBER, Brockert, BrokenSegue, Bruce1ee, Burntsauce, Bwithh, CDN99, Cacycle, Caesura, Cafzal, Caltas, Camembert, Can't sleep, clown will eat me, CanDo, Cdogsimmons, Cenarium, Centrx, Century0, Charles Matthews, ChemGardener, Chemistrypal, Chillum, Chris the speller, Chrislk02, Christian List, Christopher Parham, Chun-hian, Ck lostsword, Clackmannanshireman, Clemwang, CobraWiki, Codex Sinaiticus, Coinpeice, Colonies Chris, Computerjoe, Conversion script, Cronullasoutherland, Crystallina, Curps, Cyktsui, Cyw90, DARTH SIDIOUS 2, DMacks, DVD R W, Daniel5127, Danny, Danski14, Dantecat, Darenshan54, DarkCatalyst, Darrien, Darthgriz98, Daven200520, David Little, David Stapleton, DavidLevinson, DavidRader, Dawn1234, Dcoetzee, Delirium, Deltabeignet, Demophon, Denali134, Depressio, DerHexer, Derek.cashman, Deryck Chan, Desertsky85451, Deviator13, Devnull17, Dick slicky, DigitalGhost, Dinosaur puppy, Djcozmik, Doctormatt, Docu, Dodger67, Dodgerdave, Dogposter, Doug Alford, DrBob, Dreg743, Dren55, Dtgm, Dustimagic, Dylan Lake, ESkog, EamonnPKeane, Eaolson, EarthPerson, EdChem, Edgar181, Edonovan, Edwy, Egil, El C, ElinorD, EllaRess, Emojones, Emperorbma, Enormousdude, Eric Forste, EscapingLife, Eugenedakin, Evercat, Evil saltine, Explicit, Extra999, Ezeu, F2020, FF2010, Fahlmanb, Falcorian, Fatdemon, Fdevilbis, Feezo, Feydakin, Fiestabowl 2003, Figure, Filefire, Fireyair, Flockmeal, Flubeca, Francs2000, Frecklefoot, FreplySpang, Fte, Fuzheado, Fuzzywallaby, Fys, GDonato, Gadfium, Galoubet, Gareth Owen, Gashead 12, Gentgeen, Ghutchis, Giftlite, Gilligan mark, Girdi, Glass Sword, Glossary, Gmhisham, Gnangarra, GnuDoyng, Go for it!, Gossip14, Graham87, GregorB, Greyhood, Grstain, Gryphn, Guanaco, Gunmetal, Gurch, Gwernol, HJ Mitchell, Hadal, Hajhouse, Hallenrm, HansHermans, HappyApple, HappyCamper, Hattar393, Hdt83, Headbomb, Henry W. Schmitt, Hephaestos, Historyfan04, Historyfan40, Hotdog123, Hqb, Hraefen, Hs0225, Hugo-cs, Hummer0800, Hurricane Angel, Husond, II MusLiM HyBRiD II, IRWolfie-, Ian Pitchford, Ianblair23, Icairns, Ignitus, Ilinardi, Immortal301, Innotata, Intelati, Introgressive, Io anthony, Iranway, Iridescent, Iridium77, Isiz, Island, It Is Me Here, Ithacagorges, Itsokcuzzimme, Itub, IvaB, Ixfd64, J Di, J. Spencer, J0taj0ta, J8079s, JDPhD, JForget, JWidjaja, JaGa, Jagged 85, James086, JamesBWatson, Javit, Jawz, Jayron32, Jcbutler, Jd323, Jeeny, JesseW, Jhafkjdasfhkjalsdfh, Jian Heseri, Jiddisch, Jimothytrotter, Jklin, Jmlk17, Jmw0000, JoanneB, John, John Fader, JohnOwens, JohnT, JohnVbs, Johnbrownsbody, Johnny1926, Jose Ramos, Jose77, Joshbuddy, Jossi, Jrockley, Js11, Julesd, Jumbuck, Jurema Oliveira, Justpetehere, Kablammo, Kandar, Karam.Anthony.K, Karlwick, Karnesky, Karol Langner, Kasparov, Katzmik, Keithaber, Kelapstick, Ken Gallager, Ken2, Kesegich, Kevin Ryde, Killerkoz, Kilo-Lima, Klenje, Kntrabssi, Kokoriko, Koyaanis Qatsi, Kpjas, Kross, Kumioko, Kungfuadam, Kungming2, Kuru, Kurykh, KyleRGiggs, L Kensington, La goutte de pluie, Lakers, Lanternix, Larry_Sanger, Lcarsdata, Lectonar, Leeman056, Lepps67, Leszek Jańczuk, Levineps, Lexor, Lfh, LifeStar, Liftarn, Llull, Locos epraix, Loffelymistayke, Logologist, Looxix, LuigiManiac, Luk, LukeSurl, Luna Santin, M stone, M-le-mot-dit, MADDOX456, MER-C, MP123, Mac, Magister Mathematicae, Magore, Majorly, Malbi, Mandarax, Manishtulijnv, Maples1212, MarSch, MarcoTolo, Markan, Martinp23, Masterchamp2, Masterjamie, Masterpiece2000, Materialscientist, Matt0d, MattieTK, Mattyh190, Maurreen, Mausy5043, Mav, Maxim, Maximus Rex, Maxkirk1, Mayumashu, McDogm, Mdd, Merylholl, Meta, Mets501, Mhking, Mic, Michael Hardy, Michele123, Micmic28, Mikael Häggström, Mike Rosoft, Mike133742, Mindunder36, Minna Sora no Shita, Misza13, Modular, Mr blo0bby, Msn26586, Muchness, Mwanner, Mxn, Myepraine, Mysdaao, N2e, Nabla, Nakon, Nanoman657, Narco, Natalie Erin, NatureA16, NawlinWiki, NeallRC, Neem11, Neffyring, Netoholic, Never give in, NewEnglandYankee, Nick Y., Nihiltres, Nikai, Nikhil s n, Nikola Smolenski, Nima1024, Nitya Dharma, Nivix, Noetica, Nono64, Nov ialiste, Nposs, Nukeless, Numbo3, Obssession, Ohnoitsjamie, Oleg Alexandrov, Olivier, Omoo, Onorem, Onzenus, Opelio, OverSS, Owlly1, Oxymoron83, P99am, Padillah, PaePae, Pakaran, Palica, Paracit, Passw0rd, PatPeter, Paul August, Paul EJ King, Paulmcdonald, Pb30, Pengo, Pentasyllabic, Persian Poet Gal, Peruvianllama, Peter, Pfffff, Pgk, Philippe, Philipwgustafson@msn.com, Physical Chemist, Physicsformorons, Pie Man 360, Pigman, Pipedreamergrey, Pizzaflav, Pking123, Planetary Chaos, Plastikspork, Plato, Pmlineditor, Poetaris, Polimerek, Politepunk, Postoak, PseudoSudo, Punanypounder619, Quadell, Quintote, Quinwound, Qutezuce, Quux, Qxz, R'n'B, R6MaY89, RTC, RYNORT, Ralesk, Ran, Rashad9607, Raven4x4x, Rayquang, Rdsmith4, RebelBodhi, Red Winged Duck, RedCoat10, RekishiEJ, Remi0o, René Irwin, RexNL, Riana, Rich Farmbrough, Richard L. Peterson, Richard001, RichardF, RickK, Rmhermen, Rogianos, Romanm, Rootybeer4eva, Roozbeh, Rpresser, Rpyle731, Rune.welsh, RyanJones, Ryanleyland, Ryanminier, Ryt, Ryulong, S, SDC, SFC9394, SU Linguist, Sadi Carnot, SailorMan1863, Sakimonk, Salsa Shark, Salsb, Sam Korn, Samlyn.josfyn, SandyGeorgia, Sankalpdravid, Sannse, Savant13, Sbrools, Sbrown411, Schzmo, Scientific29, Scm83x, Scottydude, Scwlong, SebastianHelm, Secfan, SecretAgentMan00, Seidenstud, Seijihyouronka, Semper discens, Sengkang, Seth Ilys, Sgodth, Shanata, Shanes, Shellreef, Shenme, Shimgray, Shimmeringtilly, Shinjiman, Shlomke, Shootbamboo, Shultzc, SidP, Sikkema, Silverturtle1, SimonP, SimpsonGrl, SineWave, Sjakkalle, Skixz, Slowking Man, Smack, Smallman12q, Smeira, SmilesALot, Smtaylor, Snisfrealaaa, Sodium, Soilguy2, Soilguy3, Soliguy, Soliloquy, Solipsist, Sopoforic, Soumyasch, Spacepotato, Spliffy, Sploonie, SpookyMulder, Spudst3r, SpuriousQ, Squids and Chips, Srice13, St.daniel, Stan J Klimas, Stan Shebs, Steinbach, Stephen Gilbert, Stephenb, Steve Quinn, Steveking 89, Steveohare, Stevertigo, StevieNic, Stewartadcock, Stitches2021, Stwalkerster, SuPeRnAtUrAlLoVeR, SudhirP, Suisui, SuperNitro, Supermaster2011, Syncategoremata, Takometer, Talrias, Tannin, Targetter, Tarquin, Taz hunt, Tedder, Tedernst, Teenage wedding and, Template namespace initialisation script, Terebigemuwan, Texture, Thabhab, The Original Wildbear, The Rambling Man, The Thing That Should Not Be, The Transhumanist, The Transhumanist (AWB), The wub, TheBendster, TheKMan, TheLegendLions, ThePointblank, TheRanger, Thechamelon, Theda, Theislikerice, Themaab, Therefore, Theresa knott, Theshibboleth, Tigershrike, Tim Starling, TimNelson, Timir2, Timmothias, Timwi, Tiptoety, Tjayh913, Tmcsheery, Tobby72, Tom harrison, Tomasz Prochownik, Tomaxer, Tomgally, Tone, Tpbradbury, Tpth, TracySurya, TransientAircraft, Trevor MacInnis, Turlo Lomon, Tyler-07, UberScienceNerd, Ucanlookitup, Ultratomio, Under22Entreprenuer, Urania3, V8rik, VDWI2, Vaultdoor, Velblod, Velv, Victorbatista, Vig vimarsh, Violetriga, Visor, Vitalii.com, Voorlandt, Votd, Vox Rationis, Vsmith, Vssun, Vulcan Hephaestus, WakiMiko, Wakka, Walkerma, Wavesmikey, Wayne Miller, Whoda, Wikidudeman, Wikieditor06, WikipedianMarlith, Wikipediausernumber1, Wimvandorst, Wingman chemie, Wonglokking, Wood Thrush, Wowzorz, WpZurp, Wshun, Wurdwiz, XJamRastafire, Xchrisblackx, YAGSIMEHC, Yahel Guhan, Yamaguchi, Yamamoto Ichiro, YellowMonkey, Yojodo, Youssefsan, Yurivict, Zachary, Zaijaj, Zakian49, Zazaban, Zeno Gantner, Zhinker, Zoicon5, ZorkFox, ~K, , 1369 anonymous edits

Analytical chemistry *Source*: http://en.wikipedia.org/w/index.php?title=Analytical_chemistry *Contributors*: 131.227.185.xxx, 1exec1, 7, APH, Academic Challenger, Acegikmo1, Achem, AdjustShift, Age3232, Ahoerstemeier, Airridi, Akubra, Alchimio, Alexander Ogorodnikov, Andre Engels, Annedbt, Asdlib, Bcstiles, Bduke, Bensaccount, Bernarddb, Biscuittin, Bmk, Bob, Bobby1011, Bobomarch, Bomac, Bongwarrior, Bueller 007, CWenger, Cacycle, Capricorn42, Carl.bunderson, Cdang, ChemGardener, Chemical Engineer, Christopherlin, Chuq, Cnilep, Conversion script, Cow109girl, Cubic Hour, Cutler, DMacks, Davidthal, Debresser, Deirdre, Dentren, Divide, Dreg743, Eleassar, Elminster Aumar, Email4mobile, Ewlyahoocom, Frankvangeel, Fuzzywallaby, Garyzx, Gemtiger, Gentgeen, Gilligan mark, GoingBatty, Graeme Bartlett, GridEpsilon, Haham hanuka, Headbomb, Icairns, Ike9898, Iridescent, Iris lorain, Itub, IvanLanin, J.delanoy, Jaeger5432, Jeff G., JoJan, John, Johnson-gray, Jokbong, Kanzure, Kevyn, Kingpin13, Kkmurray, L33tweasley, LiDaobing, LissaAvery, Littlealien182, MPerel, Macatanong08, Materialscientist, Mav, Max.zuo, Mayumashu, Meisam.fa, Mel Etitis, Michael Devore, Michel Awkal, Micrecol, Monkeynoze, Mononomic, Morgenrot42, MrOllie, Muhandes, Mxn, Mzhou45, Nick Y., Ninabhm, Nono64, Oiws, Oleg Alexandrov, PSUMark2006, Pauluskc, Peter, Peteronium, Phillip J, Piano non troppo, Qswitch426, Quaque, R, Ranveig, RekishiEJ, Remi0o, Remyrem1, Rich Farmbrough, Rifleman 82, Rjm656s, Rjwilmsi, Rror, Rubin joseph 10, ST47, Sahehco, Saxbryn, Schnolle, SemperBlotto, Sineswan, SpaceFlight89, Speedster0404, Spencer, Stickee, Svanslyck, Teamturnz, Template namespace initialisation script, Terri G, The Anome, The Thing That Should Not Be, The wub, Tide rolls, Tom harrison, TriniTriggs, Unaiza Uzair, VASANTH S.N., Vasu2008, Velella, Vsmith, Wavelength, Wayland, Wolbo, WordyGirl90, WpZurp, WriterHound, Yk Yk Yk, Јованвб, Александър, 240 anonymous edits

Physics *Source*: http://en.wikipedia.org/w/index.php?title=Physics *Contributors*: (3ucky(3all, -- April, ..Playa187.., 0, 100110100, 10metreh, 130.225.29.xxx, 168..., 1984, 1exec1, 2D, 5 albert square, 83d40m, 97parnellj, A. di M., A7x4ev7x7x7x7x, AK2AK2, APH, AV2, AVand, Aanand Pranav Sharma, Aaron Schulz, Abdullais4u, Aboeing, Academic Challenger, Accurizer, Ace56, Acroterion, Adashiel, Addshore, AdjustShift, Adventurer, Aecis, Agaudin, Ageekgal, AgentPeppermint, Agradman, Ahoerstemeier, Aidanscotch, Airumel, Aitias, Ajitflora, Ajraddatz, Ak47K, Akamad, Akriasas, Aksi great, Alan16, Alansohn, Albino Ibis2, AlefZet, Alex7796, Alexnye, Ali, Allstarecho, Alphachimp, Alt9372, Am2006, AnAj, Anarchic Fox, Anarchopedia, Ancheta Wis, Andejons, Andonic, Andre Engels, Andrewperlmutter, Andrewpmk, Andris, Andy Marchbanks, Angela, Animum, Anna Lincoln, Annetit, Anpecota, Anrie, Antandrus, Anthony22, Anticent, Antonio Lopez, Ap, AppleJordan, Apteva, Apyule, Archanamiya, Arctic Night, ArdClose, ArglebargleIV, AriGold, Aristotle Archimedes Galileo Newton Einstein, Ark, Army1987, Arroww, Art Carlson, Arvindra, Asdfgh1234567, Asdfghjklqwertyuiopo, Ash, Ashandpikachu, Ashwinuoh, Assyria 90, Atallcostsky, Athenean, Austin Maxwell, Av99, Avenged Eightfold, Avicennasis, Awadewit, Awickert, AxelBoldt, Ayserghnaim40, AzaToth, Azadehdeath, Aznmastaz, B5Fan2258, BRG, Baa, Balaam42, Ballsdeep420, BannanaBBTeracottaBTTPie, Barak Sh, Barek, Baronnet, Bass fishing physicist, Bbatsell, Bci2, Bcrowell, Bduke, Beast of traal, Beland, Belard, Ben Ben, Bennirubber, Benzado, Bequinha, Bergjosh, Besselfunctions, Bgpaulus, Bhoutika, Big Bird, BigBritt08, Bigtimepeace, Bihco, Bill Nye the wheelin' guy, BillWSmithJr, Billtheshirk, Binksternet, BirdKr, Bishop of banter, Blackynorris, Blink'em, Blooooo, BlueDevil, Bluemask, Bmeguru, Bmk, Bmromero, Bobo192, Bogdangiusca, Bollywoodfarsia, Bongwarrior, BoomerAB, Borislav, Bossmace, Brandon, Brian Crawford, Brian0918, Brianga, Brianjd, Brion VIBBER, BrokenSegue, Bryan Derksen, BryanD, Bsod2, Bucketsofg, Budster0, Burzmali, BusterD, C quest000, C.Fred, CALR, CIreland, CWY2190, CWatchman, CWii, CYD, Cab91, Caiaffa, Calabraxthis, Calccrazy1dx, Caleb4441, Callmarcus, Caltas, Camembert, Can't sleep, clown will eat me, Canadian-Bacon, Canderson7, Canis Lupus, CapitalR, Capricorn42, Cardamon, CardinalDan, Carlos177567, Carnildo, Casper2k3, Catcom101, Catgut, Cbrown1023, Cdc, Cedars, Celebere, Ceranthor, Cesaar, Cfwalther, Charles Matthews, Charley white, CharlotteWebb, Chelman, Chewings72, Chhe, Chich03, Chocolog, Cholmes75, ChrisHamburg, Christian List, Christoph Scholz, Christopher Parham, Chzz, Ciozi137, Circeus, Citicat, Ckatz, ClodiSK, Closedmouth, Cmpd1, CoasttoCoast, Cobaltcigs, Codetiger, Codex Sinaiticus, Cometstyles, CommonsDelinker, Coneslayer, Confused111, Connorcostello, Connormah, Conpul, Conversion script, Conwiktion, Cool3, Copysan, CorvetteZ51, Couki, Courcelles, Cphighkid, Craig Pemberton, Crazycomputers, Crazynas, Creidieki, Cremepuff222, Crusadeonilliteracy, Cryoboy, Csigabi, Curps, Cyber-witten, Cybercobra, Cyberied, Cyclonenim, DARTH SIDIOUS 2, DBishop1984, DJ Clayworth, DMacks, DVD R W, DVdm, Da monster under your bed, DabMachine, Dadude3320, Dalstadt, Damitr, Dan D. Ric, Dandrake, Dane Gustafun, Danger, Daniel Mahu, Daniel Quinlan, Dank, Danski14, DarkFalls, DarkRain, Darkspots, Darkwind, Darth Panda, DarthShrine, Dave001001, Dave6, Davewild, David R. Ingham, DavidLevinson, Davidclifford, Dawn Bard, Db099221, Debresser, Defender of torch, Deglr6328, Dekisugi, Demerzel, Demmy, Denholm Reynholm, Deor, DerHexer, Deryck Chan, Deskana, Dgrant, Dhc529, Diego pmc, Dina, Dipics, Discospinster, DivineAlpha, Djr32, Dmr2, Doc glasgow, Docu, Dondervogel 2, Donhalcon, Dori, Dougofborg, Doulos Christos, Downw thesickness, DragonflySixtyseven, Drdaco, Drjem3, Drmies, Drumsac, Dtgm, Dudesleeper, DumbAndDumber2000, Dumdumdom,

Dupz, Dusty669211, Dyknowsore, Dysprosia, EJF, EPM, ERK, Eager McBeaver, Ed g2s, EdBever, Edgar181, Edward, Edward Morgan Blake, Edwardo66, Eeekster, Eggwee, Egil, Egmontaz, El C, Eleassar, EllenPetersen, Elockid, Eloquence, Emarti (usurped), Emc2, Ems2715, Ems57fcva, EncycloPetey, Enormousdude, Enviroboy, Epbr123, Epic Penguin123, Erebus555, Eric Herboso, Ericnutsch, Erik9, Esanchez7587, Estarriol, Eu.stefan, EugeneZelenko, Euryalus, Evan Mayer, Evercat, Everyking, Exert, Explicit, F, F Notebook, Faus, Faustina.minor, Favonian, Fbazjohn, FeloniousMonk, Fergo100, Fibonacci, Fieari, Fieldday-sunday, Filll, Finalius, Fingerz, FiniteTree, Fl, Flashgordon12, Flauto Dolce, Flewis, Flyer22, Flyguy649, Fmiddleton, Frankenpuppy, Franklin.vp, Freakinadrian, Freakofnurture, Fred Bauder, Fredbauder, Fredrik, FreplySpang, Frito, Frosted14, Fs923, Funkydunky9, Fuzheado, Fyyer, Fæ, G. Campbell, G. Moore, GKaczinsky, Gaara unleashed, Gaara144, Gabbe, Gabi bart, Gabrielleitao, Gahs, Gail, Galaxiaad, Galluer, Galoubet, Garion96, Gauze123, General Galavan, General Reader, Geobio, Geometry guy, Georgelikesboys, Georgelulu, Ghaly, Ghurtad1, Ghyll, Giftlite, Gilliam, Gina va gina va, Gioto, Giraffedata, Gjd001, Glane23, GlasGhost, Gnixon, Go for it!, Gogo Dodo, Golbez, GoldenMeadows, Goldom, Goozle, Grafen, Graham87, Graysen98, Greenguy1090, Grubber, Grunt, Grunty Thraveswain, Gtg204y, Gunslinger of Gilead, Gurch, Gurchzilla, Guérin Nicolas, Gwernol, HKT, Hadal, Haisch, HalfShadow, Haljolad, Harald88, Harryboyles, Haselhurst, Haus, Hcollin, Hdt83, Headbomb, Hellboy2hell, HelpInHandy, Herakles01, Herbee, Hermione1980, Heron, Hidaspal, Hike395, Holbergj, Hollyellis, Horselover Frost, Howcheng, Hrishikesh911, Huangdi, Hugh Mason, Hungrypoo, Huppafa, Husond, Huwwaters, Hydraton31, Hydrogen Iodide, I already forgot, II MusLiM HyBRiD II, IIsL337, IRP, Icairns, Ice Cold Beer, Ilmari Karonen, ImGladMyMomIsDead, Imadeausrname, Immualkurt, Immunize, ImperfectlyInformed, Imran, IncognitoErgoSum, Insvik, InvertRect, Inyuki, Ipatrol, Iridescent, Ironholds, Isis, Island, Isomorphic, Itub, Ixfd64, J.delanoy, J04n, JDCMAN, JForget, JFreeman, JHBrewer, JHMM13, JJ Harrison, JNW, JRGBruno, JYOuyang, JaGa, JabberWok, Jackelfive, Jackol, JackyR, Jade Knight, Jagged 85, Jake Wartenberg, Jamie fearon, JamieS93, Javainone, Jaydawg89, Jbarta, Jbergquist, Jeff G., Jehochman, Jeronimo, JessBr, Jfdwolff, Jim E. Black, Jim woz ere, Jim62sch, Jimothytrotter, Jitse Niesen, Jjasi, Jklin, Jmabel, Jmundo, Jni, Joanee Woolee, JoanneB, JoeBlogsDord, John, John254, JohnCD, JohnOwens, Joke137, Jon Awbrey, Jon513, Jonie2102, Jorend, Jose77, Josh the Nerd, Jossi, Joymmart, Jredmond, Jsc83, Julesd, Juliecoolgirl34, Jumping cheese, Jusdafax, Jusjih, Just Call Me Einstein, Justinfr, Justup2002, K, KGasso, KLHS6, Kaihsu, Karada, Karen Holman, Karlthegreat, Karlwick, Karol Langner, Kasparov, Katalaveno, Katieh5584, Kazvorpal, Kbh3rd, Keegan, Keesiewonder, Keith-07, Kelvin Case, Keo, KerathFreeman, Kevin B12, Kevin aylward, Kf4bdy, Khalid Mahmood, Kiba, King of Hearts, KingTT, Kingpin13, Kmizzledabombdizzle, KnowledgeOfSelf, Knulclunk, Kornfan71, Koyaanis Qatsi, Koyn, KrakatoaKatie, Krea, Kribbeh, Kubigula, Kungfuadam, KurtRaschke, Kuru, Kushal one, Kville105125, Kwekubo, Kwsn, Kyoko, LAX, LaRouxEMP, Lagrangian, Lambertian, Lantonov, Laplace's Demon, LaszloWalrus, Laurascudder, LaurenceBoss, Laverty-j2, Laxragaza, Lcarsdata, Leafyplant, LeaveSleaves, LedgendGamer, Lee Daniel Crocker, Lenticel, Leobyun, Levineps, LexCorp, Lexor, Lightmouse, Lights, Lisatwo, Little guru, Little_guru, Littlealien182, Livilou517, Llull, Logan, Logologist, Lokal Profil, Loodog, Loom91, Looxix, Lotje, Lseixas, Lumidek, Luna Santin, Lupin, Lupo, Lycurgus, M, M C Y 1008, M-le-mot-dit, M.thoriyan, MASQUERAID, MC10, MER-C, MLauba, MONGO, MPerel, Maddemo, Madesfuga, Madhero88, Madman, Maganjeanthrower, Magic5227, Magister Mathematicae, Magnumstealth, Magnus Manske, Magog the Ogre, Makewater, Malafaya, Malhonen, Mannafredo, MarSch, Marijuanaboy, Markan, Markashley1990, Marnanel, MarsRover, Marsrocks09, Martin Hogbin, Masamunecyrus, MassimoAr, MastCell, Master shepherd, Materialscientist, Matthead, MattjenningsXxXxXD, Mattmac, Mattyisafatty, Maurice Carbonaro, Maurreen, Mav, MaxSem, Maximus Rex, Maxis ftw, Mayumashu, McSly, McVities, Mdd, MearsMan, Meelar, MegaSloth, Melchoir, Men99999, Meno25, MercuryBlue, Methcub, Mets501, MetsFan76, Mic, Michael C Price, Michael Hardy, Michael K. Edwards, MichaelMaggs, Miffy900, Mike Peel, Mike Rosoft, Mikeblas, Miketwardos, Minesweeper, Minna Sora no Shita, Mista poe, Mithent, Moink, Molobo, Moncrief, Monedula, Moni3, Monkeykiss, MoogleEXE, Moriori, Morwen, Mpatel, Mr Stephen, Mr. Billion, Mr. Know-It-All, Mr. Lefty, Mr.Nostalgia, MrOllie, Mschel, Mtruch, Muchness, Murdajvc, Murtasa, Mushroom, Mwanner, Mwborwmb, Mxn, Mynameisearl12345, N5iln, NJA, Najoj, Nakon, Naktys, NarSakSasLee, Narutohaku, Navasj, Naveen Sankar, Naveensankar, Naveeyindren Thangrajah, Navilon, NawlinWiki, Nburden, Neelix, NeilN, Nein, Neonstarlight, Netoholic, Never give in, NewEnglandYankee, Newportm, Nickmanning214, Nicolas wolfwood, Nigholith, Nihiltres, Nikola Smolenski, Nimocks, Nineg, Nishantnag1992, Niteowlneils, Nitya Dharma, Nivix, Nixdorf, Njaelkies Lea, Nnedass, No Guru, Noctibus, NodnarbLlad, Noeckel, Noetica, Notinasnaid, Notjake13, Notreallydavid, Nowrunningweb, Nsaa, Nuclear ace, Nufy8, Nvf, Nygiants2744, Nyh, O process, OBI EMMANUEL, Odd nature, Odie5533, Ohnoitsjamie, Oisuzie, Ojay123, Old Moonraker, Oleg Alexandrov, Olgaflow13, Oli Filth, Olivier, Olleicua, Omicronpersei8, Onceonthisisland, One Night In Hackney, Onestone, Orangemarlin, Orion11M87, Orionix, Ottershrew, OverSS, Oxymoron83, Ozcep, PL290, PSU PHYS514 F06, PaddyLeahy, Pandanator75, Panoramix, Panplenty, Panser Born, Papa November, Passandid, Passw0rd, Patilsagar09, Paul August, Paulmcdonald, Pcarbonn, Pelivani, Pennie666, Pentasyllabic, Penubag, Persian Poet Gal, Petiatil, Petitspois, Pgk, Phaedriel, Pharaoh of the Wizards, Pharos, Phdkimd, Phgao, Philip Trueman, Phisicsfa, PhotoBox, Phusis, PhySusie, Phys, Physics - The Science of Nature, Physicsformorons, Pierpontpaul2351, PierreAbbat, Pilotguy, Pinethicket, Pinky raj01, Piotrek54321, Pjoef, Pjvpjv, Plastikspork, Plumbago, Pmlineditor, Poeloq, Polluxian, Polypipe Wrangler, Polyvios, Possum, Prestonhenry, PrimeCupEevee, Przepla, Psyche825, Ptk, Purplees2, Qorilla, Quadell, Quiddity, Quintessent, Qutezuce, Qwyrxian, Qxz, R'n'B, RG2, RJHall, RJaguar3, RTC, Rabarberski, Radhisure, Radon210, Rahulvohra29, Railer 295, Ramir, Randomblue, Rasmus Faber, Raul654, Raven4x4x, Ray765, Razism99, Razorflame, Rbeas, Rbj, Rdsmith4, Reach Out to the Truth, Reddi, Reedy, Reerpizza, Reliableforever, ReluctantPhilosopher, RenamedUser2, Rettetast, Revilowaldow, RexNL, Rexc2011, Rhrad, Riana, Rich257, RichardF, Richieisking19, Rickyrossdaboss, Rightly, Rise Against713, Rising Dragon Fist, Rivertorch, Rl, Rls, Rmalloy, Rmrfstar, Roachie, Rob Mahurin, Rocks on the road, Rockzer, Roflbomb, Rogwan, Roo72, Rootybeer4eva, Rory096, Rotem Dan, Rracecarr, Rror, Rtyq2, RunOrDie, Ruud Koot, Ryan Roos, RyanGerbil10, Ryeterrell, Ryugahideki, Ryulong, SCZenz, SEWilco, SFC9394, SJP, ST47, Sadi Carnot, Saeed.Veradi, Saganaki-, Sai2020, Saikiri, Salmar, Salsa Shark, Salsb, Sam Coutu-Oughton, Sam Korn, Sam Li, Samlyn.josfyn, Samoboow, Samsonov1947, Sandeepsreevastav, SandyGeorgia, Sango123, Santaisnotyourfriend, Saros136, SaveThePoint, Sceptre, SchfiftyThree, Schroer, Schzmo, Sciencenerdgeek, Scott McNay, Scott3, Seatdistrict, Sengkang, Serapio, Shadowbane1001, Shadowoftime, Shalom Yechiel, Sharkface217, Sharmarkeali, Shawn in Montreal, Shebeer, Shell Kinney, Shepherrd, Shesterov, Shoeofdeath, Shoessss, Sietse Snel, Sifaka, Sign Creator, Silambarasan Physics, Silence, Silly rabbit, Silverlining5678, Simi 11 krish, Simon009988, SimonP, SiobhanHansa, Sionus, Sir Nicholas de Mimsy-Porpington, SivaKumar, Sjakkalle, Sjking92, Sjö, SkerHawx, Skeye1215, SlamDiego, SlashDot, Sljaxon, Slmader, Slowking Man, Smenge32, Snowolf, Snoyes, SoSaysChappy, Some Color Mage, SomeStranger, Someone else, Soumyasch, Spades92, Spiff, Spike Wilbury, SpikeToronto, Spitfire8520, Sputnikcccp, Squishey, Srleffler, Srnec, Staffwaterboy, Standardfact, Starkrm, Starship Trooper, StaticGull, Staxringold, Stefan Jansen, Stefo, Stephenb, Steven Zhang, Stevenwagner, Stevertigo, Straighter, Stsang, Stultus, Sturmde, Sultangris, Sum1cool, Sundar, Sunray, Supersion77, Support.and.Defend, Susurrus, Svanslyck, Sverdrup, Sylv3st3r101, T-borg, THEN WHO WAS PHONE?, TML, TSchwab, TTE, Ta bu shi da yu, TachyonP, Tadas12, Taejin, Talyor Will, Tamjid, Tankmaster01, Tannin, Tanvir Ahmmed, Tarotcards, Tarquin, Tasudrty, Tavilis, Tawnyman1, Tbhotch, Tcncv, TedE, Teh Rote, Template namespace initialisation script, Terfili, Tevatron, Tgv8925, Thaejas, That Guy, From That Show!, The Anome, The High Fin Sperm Whale, The Original Wildbear, The Psycho, The Thing That Should Not Be, The Transhumanist, The Transhumanist (AWB), TheFeds, TheFox3, Thehamman55, Thehotelambush, Theirishpianist, Thenamenottaken, Theoneintraining, Thepretender41, Theshibboleth, Theunixgeek, Thewayforward, Thingg, This user has left wikipedia, Thomaswgc, Tide rolls, TigerShark, Tigershrike, Til Eulenspiegel, Tim Shuba, Tim Starling, TimNelson, TimothyRias, Timwi, Titoxd, Tjlafave, Tmac88ivrsn, Tmsmith, Tom harrison, Tomasz Prochownik, Tombomp, Tommy2010, Toohool, TracySurya, Trafford09, Traroth, Trelvis, Trevor MacInnis, Trevorcavell, TriTertButoxy, Triwbe, Tronkete, Troy 07, Trulystand700, Trumpetplaya44, Trusilver, Truthnlove, Tspilman, Tsumetai, Ttlxxx, TutterMouse, Tyler, Typuifre, Tyrol5, Tytrain, U.S.A.U.S.A.U.S.A., UberCryxic, Ukulele, Ultimus, Uncle Dick, Unigfjkl, Unknown123123123, Urgos, Urvabara, Uugedsaz, VMS Mosaic, Vampus, Van helsing, Vanished User 1004, Vanished User 4517, VanishedUser314159, Vary, Vatoking, Vedant lath, Velblod, Velella, Velvetron, Versus22, Vikasmorankar, Vinay tv, Vincenzo.romano, Vipinhari, Voigfdsa, Voyajer, Vsmith, Vssun, Vulcan Hephaestus, WPjcm, Waggers, Wakka, Walkerma, WanderingFinches, Wardigong, Washable20, Wavelength, Wayward, Wdfarmer, Wdjunkin, WebHamster, Webmeischda, West Brom 4ever, Whassuo, White Cat, WhiteC, Whiten667, Wiki Administrator of Physics, Wiki alf, Wiki woo, WikiSlasher, WikiWacked, Wikid77, Wikiklrsc, Wikimichael22, Wikitumnus, Wiktionary4Prez!, Wildnox, Wile E. Heresiarch, William M. Connolley, Wimt, Winhunter, Winston365, WinterSpw, Wintonian, Wisebridge, Witten Is God, Wizardman, Wolfkeeper, Wolfmankurd, Wolfrock, Wood Thrush, Woohookitty, Wronkiew, Wtmitchell, XJamRastafire, Xemoi, Xezbeth, YK Times, Yahel Guhan, Yaley101, Yamamoto Ichiro, Yanksox, YassineMrabet, YellowMonkey, Yes, I'm A Scientist, Yevgeny Kats, Yidisheryid, Yong Liang, Youssefsan, Yugolervan, YusrSehl, Zac wise, Zachary, Zachorious, Zakian49, Zarniwoot, Zazaban, ZimZalaBim, Zoicon5, ZomBGolth, ZsigE, Zsinj, Zsynopsis, Zundark, Zvis, Zzuuzz, ^demon, Тне ежесабботочный, , , 3351 anonymous edits

Groundwater *Source*: http://en.wikipedia.org/w/index.php?title=Groundwater *Contributors*: A. B., Aharlan, Aitias, Alansohn, Ali K, Angela, Anlace, Antandrus, Anthonares, Anthony Appleyard, Aoi, Arakunem, ArcticFrog, Argyriou, Arthena, Badgernet, Barek, Basar, Bayerischermann, Beetstra, Beland, Bigbluefish, Bobo The Ninja, Bobo192, Bryan-Oddfellow, CDM2, CSWarren, Calltech, Caltas, Calvin 1998, CarolGray, Ccgrimm, Charles Dickens, Chicco3, Chriswaterguy, Chroja, Chuunen Baka, Civil Engineer III, Clayton orgles, Clifftreyens, Cmdrjameson, Cp111, Cwiweb, DJKostaki, DVD R W, Daniel Collins, Dariusofthedark, David Levy, Dawn Bard, DeansFA, Discospinster, Djdutch, Doniago, Drm310, Dzubint, Eastlaw, Elfino, Ellisthion, Emc2, Entropy, Epbr123, Epipelagic, ErikDunsing, Everyking, EvilPizza, Fabiform, FayssalF, Fieldday-sunday, Fiveless, Forstbirdo, Fourdee, Frankenpuppy, Frazzydee, FreplySpang, Fresno2000, Gcm, Gdphdb, GentlemanGhost, Geographer, Geologyguy, Glen, Glenn, Gob Lofa, Gralo, GregE625, Gunnernett, Hdt83, Hollowtools, Huzzahmaster018, Indiawaterportal, Infrangible, J.delanoy, J04n, JForget, JTN, JamesBWatson, Jarbon, Jauerback, Jeff G., Jessa marie, JodyB, John254, Johnathonsmithe, Jon Nevill, Jrdioko, Julesd, Jusdafax, Jóna Þórunn, Kateshortforbob, KatieDOM, Katoa, Kbh3rd, Khalid hassani, Koiplaats, Krellis, KrisK, Ksyrie, Kuhlman, Kungming2, Kwells1989, L Kensington, Leafyplant, Lilac Soul, Loltamere, Look2See1, Mac, Madhero88, Mahjongg, Mailseth, Malo, Marisa269x, Marshman, Martin451, Materialscientist, Maurreen, Megiddo1013, Mentifisto, Mikenorton, Mintleaf, Mirv, Mnemo, Moeron, Monfornot, Monsoon Waves, Moreau1, Mygerardromance, Nakon, Natalie Erin, NawlinWiki, NcRomance, NellieBly, Nick, Nick123, Omicronpersei8, Omnicog, Otisjimmy1, Oxbox, Oxymoron83, Pacula, Panyd, Parutakupiu, Pearle, Perchedaquifer, Persian Poet Gal, Phantomsteve, Pietrow, Pill0031, Pinethicket, Pinky sl, Pollinator, Pupster21, Qfl247, Quintote, Qxz, R.J.Oosterbaan, R.steenhard, R123triplehhhfan123, Raven4x4x, RedWolf, Redvers, Reinyday, Rjwilmsi, RocketRoger, Salgueiro, Salmar, Samw, Sean.hoyland, Seaphoto, Shiftchange, Shizane, Shrumster, Siim, Skater, Slawojarek, Smack, Smallverm, Some jerk on the Internet, Sonett72, SpaceFlight89, SparrowsWing, Steven Weston, Suntag, Syrthiss, TARATHILIEN, Tellyaddict, The Anome, The High Fin Sperm Whale, The Thing That Should Not Be, Theo Pardilla, Tide rolls, Tivedshambo, Tom harrison, Tschel, USchick, Ubergeekguy, V.narsikar, Vervin, Vincentcayabyab, Viridae, Voice of reason993, Vsmith, WadeSimMiser, WarthogDemon, Watertruth, Wavelength, Weetoddid, WikiLaurent, Wikipelli, Wilson44691, Wimt, Wrp103, Xiahou, Xionbox, Zeizmic, Zuejay, 578 anonymous edits

Petrography *Source*: http://en.wikipedia.org/w/index.php?title=Petrography *Contributors*: A Karley, Al Pereira, CranstonShenir, DanMS, Drench, Eduardoestima, Eras-mus, Gaius Cornelius, His Manliness, Hmilgram, Icairns, Jef-Infojef, Kjkolb, Martino3, Mikenorton, Mwtoews, Plantlucy, Plato, Qfl247, RedWolf, SEWilco, Shinchey, Siim, Vegaswikian, Viv Hamilton, Volcanoguy, Vsmith, Wiggin15, Zephalis, Zigurat, 15 anonymous edits

Economic geology *Source*: http://en.wikipedia.org/w/index.php?title=Economic_geology *Contributors*: APH, Bobo192, D6, DuncanHill, Dungodung, Edward, Fama Clamosa, Ivan Bajlo, IvanLanin, Ksyrie, Ligar, Michael Hardy, Mikenorton, NatureA16, Pmetzger, Rayan9807, Rolinator, Vsmith, Wik, Wilchett, Wizard191, Zoicon5, 28 anonymous edits

Image Sources, Licenses and Contributors

File:Table of Natural History, Cyclopaedia, Volume 2.jpg *Source*: http://en.wikipedia.org/w/index.php?title=File:Table_of_Natural_History,_Cyclopaedia,_Volume_2.jpg *License*: Public Domain *Contributors*: Brian0918, LaosLos, Tmizuk, Wst

Image:Buffon 1707-1788.jpg *Source*: http://en.wikipedia.org/w/index.php?title=File:Buffon_1707-1788.jpg *License*: Public Domain *Contributors*: François-Hubert Drouais

Image:Carl Spitzweg 025.jpg *Source*: http://en.wikipedia.org/w/index.php?title=File:Carl_Spitzweg_025.jpg *License*: Public Domain *Contributors*: Atamari, Bukk, Mattes, W!B:, 1 anonymous edits

Image:MakhteshGadolJurassicRidge.JPG *Source*: http://en.wikipedia.org/w/index.php?title=File:MakhteshGadolJurassicRidge.JPG *License*: Public Domain *Contributors*: User:Wilson44691

Image:tephrochronology_iceland.JPG *Source*: http://en.wikipedia.org/w/index.php?title=File:Tephrochronology_iceland.JPG *License*: GNU Free Documentation License *Contributors*: User:Dentren

File:Uz geology fieldschool2002.jpg *Source*: http://en.wikipedia.org/w/index.php?title=File:Uz_geology_fieldschool2002.jpg *License*: Public Domain *Contributors*: User:Babakathy

Image:Mudlogging.JPG *Source*: http://en.wikipedia.org/w/index.php?title=File:Mudlogging.JPG *License*: GNU Free Documentation License *Contributors*: User:Citypeek

File:Geologists-tools hg.jpg *Source*: http://en.wikipedia.org/w/index.php?title=File:Geologists-tools_hg.jpg *License*: Creative Commons Attribution 3.0 *Contributors*: User:Hgrobe

File:CratersOfTheMoonFordTent.jpg *Source*: http://en.wikipedia.org/w/index.php?title=File:CratersOfTheMoonFordTent.jpg *License*: Public Domain *Contributors*: O.E. Meinzer

Image:Classroom at a seconday school in Pendembu Sierra Leone.jpg *Source*: http://en.wikipedia.org/w/index.php?title=File:Classroom_at_a_seconday_school_in_Pendembu_Sierra_Leone.jpg *License*: Public Domain *Contributors*: L. Lartigue

File:Jewish Children with their Teacher in Samarkand cropped.jpg *Source*: http://en.wikipedia.org/w/index.php?title=File:Jewish_Children_with_their_Teacher_in_Samarkand_cropped.jpg *License*: unknown *Contributors*: , digital rendering for the Library of Congress by Walter Frankhauser / WalterStudio

File:Adriaen van Ostade 007.jpg *Source*: http://en.wikipedia.org/w/index.php?title=File:Adriaen_van_Ostade_007.jpg *License*: Public Domain *Contributors*: Emijrp, Goldfritha, Jastrow, Mattes, Vincent Steenberg, Wst

File:Teacher in Laos.jpg *Source*: http://en.wikipedia.org/w/index.php?title=File:Teacher_in_Laos.jpg *License*: Public Domain *Contributors*: User:Masae

File:Rostock Schmidt Lehrer-Student.jpg *Source*: http://en.wikipedia.org/w/index.php?title=File:Rostock_Schmidt_Lehrer-Student.jpg *License*: Creative Commons Attribution 2.5 *Contributors*: User:Schiwago

File:Bundesarchiv Bild 183-13055-0008, Hohendorf, JP mit Dorflehrer.jpg *Source*: http://en.wikipedia.org/w/index.php?title=File:Bundesarchiv_Bild_183-13055-0008,_Hohendorf,_JP_mit_Dorflehrer.jpg *License*: unknown *Contributors*: Klein

File:Koerperstrafe- MA Birkenrute.png *Source*: http://en.wikipedia.org/w/index.php?title=File:Koerperstrafe-_MA_Birkenrute.png *License*: Public Domain *Contributors*: .

File:TeacherBritishMuseum.jpg *Source*: http://en.wikipedia.org/w/index.php?title=File:TeacherBritishMuseum.jpg *License*: Creative Commons Attribution 3.0 *Contributors*: User:Etan J. Tal

Image:Chemicals in flasks.jpg *Source*: http://en.wikipedia.org/w/index.php?title=File:Chemicals_in_flasks.jpg *License*: Creative Commons Attribution 2.0 *Contributors*: Benjah-bmm27, Fjmustak, ITurtle, Ida Shaw, Karelj, Meisam, 1 anonymous edits

Image:VysokePece1.jpg *Source*: http://en.wikipedia.org/w/index.php?title=File:VysokePece1.jpg *License*: Attribution *Contributors*: Al Silonov, Darwinek, Ies, Mbdortmund, Siebrand, ŠJů

Image:Lab bench.jpg *Source*: http://en.wikipedia.org/w/index.php?title=File:Lab_bench.jpg *License*: GNU Free Documentation License *Contributors*: User:Magnus Manske

Image:Epicurus Louvre.jpg *Source*: http://en.wikipedia.org/w/index.php?title=File:Epicurus_Louvre.jpg *License*: Creative Commons Attribution-Sharealike 2.5 *Contributors*: User:Sting, User:Sting

Image:Antoine lavoisier color.jpg *Source*: http://en.wikipedia.org/w/index.php?title=File:Antoine_lavoisier_color.jpg *License*: Public Domain *Contributors*: Louis Jean Desire Delaistre, after Boilly

Image:Taxol.svg *Source*: http://en.wikipedia.org/w/index.php?title=File:Taxol.svg *License*: Public Domain *Contributors*: Calvero.

Image:Electron orbitals.svg *Source*: http://en.wikipedia.org/w/index.php?title=File:Electron_orbitals.svg *License*: Public Domain *Contributors*: user:Patricia.fidi

Image:Emission spectrum-Fe.svg *Source*: http://en.wikipedia.org/w/index.php?title=File:Emission_spectrum-Fe.svg *License*: Public Domain *Contributors*: User:Yttrium91

File:Bunsen-Kirchhoff.jpg *Source*: http://en.wikipedia.org/w/index.php?title=File:Bunsen-Kirchhoff.jpg *License*: Public Domain *Contributors*: Unkown (Courtesy Edgar Fahs Smith Memorial Collection, Department of Special Collections, University of Pennsylvania Library.)

Image:Flame test.jpg *Source*: http://en.wikipedia.org/w/index.php?title=File:Flame_test.jpg *License*: GNU Free Documentation License *Contributors*: User:WAvegetarian

Image:Analytical instrument.gif *Source*: http://en.wikipedia.org/w/index.php?title=File:Analytical_instrument.gif *License*: Creative Commons Attribution-Sharealike 3.0 *Contributors*: User:Kkmurray

File:1 MV accelerator mass spectrometer.jpg *Source*: http://en.wikipedia.org/w/index.php?title=File:1_MV_accelerator_mass_spectrometer.jpg *License*: Public Domain *Contributors*: Kkmurray, Nobunaga24

File:TLC black ink.jpg *Source*: http://en.wikipedia.org/w/index.php?title=File:TLC_black_ink.jpg *License*: GNU Free Documentation License *Contributors*: Original uploader was Natrij at en.wikipedia

File:3D-SIM-3 Prophase 3 color.jpg *Source*: http://en.wikipedia.org/w/index.php?title=File:3D-SIM-3_Prophase_3_color.jpg *License*: Creative Commons Attribution-Sharealike 3.0 *Contributors*: Lothar Schermelleh

File:Glass-microreactor-chip-micronit.jpg *Source*: http://en.wikipedia.org/w/index.php?title=File:Glass-microreactor-chip-micronit.jpg *License*: Public Domain *Contributors*: User:Micronit

Image:Calibration curve.gif *Source*: http://en.wikipedia.org/w/index.php?title=File:Calibration_curve.gif *License*: Creative Commons Attribution-Sharealike 3.0 *Contributors*: User:Kkmurray

File:Analyse thermo gravimetrique bruit.png *Source*: http://en.wikipedia.org/w/index.php?title=File:Analyse_thermo_gravimetrique_bruit.png *License*: GNU Free Documentation License *Contributors*: user:cdang

Image:Acceleration components.JPG *Source*: http://en.wikipedia.org/w/index.php?title=File:Acceleration_components.JPG *License*: Creative Commons Attribution-Sharealike 3.0 *Contributors*: User:Brews ohare

Image:Astronaut-EVA.jpg *Source*: http://en.wikipedia.org/w/index.php?title=File:Astronaut-EVA.jpg *License*: Public Domain *Contributors*: NASA

Image:Lightning in Arlington.jpg *Source*: http://en.wikipedia.org/w/index.php?title=File:Lightning_in_Arlington.jpg *License*: GNU Free Documentation License *Contributors*: User:Postdlf

Image:Pahoeoe fountain original.jpg *Source*: http://en.wikipedia.org/w/index.php?title=File:Pahoeoe_fountain_original.jpg *License*: unknown *Contributors*: Jim D. Griggs, HVO (USGS) staff photographer

Image:GodfreyKneller-IsaacNewton-1689.jpg *Source*: http://en.wikipedia.org/w/index.php?title=File:GodfreyKneller-IsaacNewton-1689.jpg *License*: Public Domain *Contributors*: Algorithme, Beyond My Ken, Bjankuloski06en, Grenavitar, Infrogmation, Kelson, Kilom691, Porao, Saperaud, Semnoz, Siebrand, Sparkit, Thomas Gun, Wknight94, Wst, Zaphod, 4 anonymous edits

Image:Bose Einstein condensate.png *Source*: http://en.wikipedia.org/w/index.php?title=File:Bose_Einstein_condensate.png *License*: Public Domain *Contributors*: NIST/JILA/CU-Boulder

Image:CMS Higgs-event.jpg *Source*: http://en.wikipedia.org/w/index.php?title=File:CMS_Higgs-event.jpg *License*: unknown *Contributors*: Lucas Taylor

Image:Hubble ultra deep field high rez edit1.jpg *Source*: http://en.wikipedia.org/w/index.php?title=File:Hubble_ultra_deep_field_high_rez_edit1.jpg *License*: unknown *Contributors*: User:Noodle snacks

Image:Modernphysicsfields.svg *Source*: http://en.wikipedia.org/w/index.php?title=File:Modernphysicsfields.svg *License*: GNU Free Documentation License *Contributors*: F l a n k e r, Karelj, Mdd, Quibik, Schekinov Alexey Victorovich, YassineMrabet, 1 anonymous edits

Image:IMG 1729 Gemaal met schroef van Archimedes bij Kinderdijk.JPG *Source*: http://en.wikipedia.org/w/index.php?title=File:IMG_1729_Gemaal_met_schroef_van_Archimedes_bij_Kinderdijk.JPG *License*: Creative Commons Attribution-Sharealike 2.5 *Contributors*: Abdullah Köroğlu, Ellywa, Fransvannes, Havang(nl), Juiced lemon, Koba-chan, WikipediaMaster, 3 anonymous edits

Image:Feynman'sDiagram.JPG *Source*: http://en.wikipedia.org/w/index.php?title=File:Feynman'sDiagram.JPG *License*: Creative Commons Attribution-Sharealike 2.5 *Contributors*: Original uploader was Ancheta Wis at en.wikipedia

Image:Meissner effect p1390048.jpg *Source*: http://en.wikipedia.org/w/index.php?title=File:Meissner_effect_p1390048.jpg *License*: Creative Commons Attribution-Sharealike 2.5 *Contributors*: Mai-Linh Doan

File:Shipot.jpg *Source*: http://en.wikipedia.org/w/index.php?title=File:Shipot.jpg *License*: GNU Free Documentation License *Contributors*: User:USchick

Image:AlapahaRiver2002.jpg *Source*: http://en.wikipedia.org/w/index.php?title=File:AlapahaRiver2002.jpg *License*: Public Domain *Contributors*: Stewart Tomlinson, Florida

File:High plains fresh groundwater usage 2000.svg *Source*: http://en.wikipedia.org/w/index.php?title=File:High_plains_fresh_groundwater_usage_2000.svg *License*: Creative Commons Attribution-Sharealike 3.0 *Contributors*: User:Kbh3rd

File:Groundwater flow.png *Source*: http://en.wikipedia.org/w/index.php?title=File:Groundwater_flow.png *License*: unknown *Contributors*: Original uploader was KrisK at en.wikipedia

Image:MiddleSpring.JPG *Source*: http://en.wikipedia.org/w/index.php?title=File:MiddleSpring.JPG *License*: GNU Free Documentation License *Contributors*: User:Citypeek

Image:Limestone building with pollution.jpg *Source*: http://en.wikipedia.org/w/index.php?title=File:Limestone_building_with_pollution.jpg *License*: GNU Free Documentation License *Contributors*: Ali K, J.-H. Janßen, 1 anonymous edits

File:LvMS-Lvm.jpg *Source*: http://en.wikipedia.org/w/index.php?title=File:LvMS-Lvm.jpg *License*: GNU Free Documentation License *Contributors*: User:Qfl247

License

If the Cover Text requirement of section 3 is applicable to these copies of the Document, then if the Document is less than one half of the entire aggregate, the Document's Cover Texts may be placed on covers that bracket the Document within the aggregate, or the electronic equivalent of covers if the Document is in electronic form. Otherwise they must appear on printed covers that bracket the whole aggregate.

8. TRANSLATION

Translation is considered a kind of modification, so you may distribute translations of the Document under the terms of section 4. Replacing Invariant Sections with translations requires special permission from their copyright holders, but you may include translations of some or all Invariant Sections in addition to the original versions of these Invariant Sections. You may include a translation of this License, and all the license notices in the Document, and any Warranty Disclaimers, provided that you also include the original English version of this License and the original versions of those notices and disclaimers. In case of a disagreement between the translation and the original version of this License or a notice or disclaimer, the original version will prevail.

If a section in the Document is Entitled "Acknowledgements", "Dedications", or "History", the requirement (section 4) to Preserve its Title (section 1) will typically require changing the actual title.

9. TERMINATION

You may not copy, modify, sublicense, or distribute the Document except as expressly provided under this License. Any attempt otherwise to copy, modify, sublicense, or distribute it is void, and will automatically terminate your rights under this License.

However, if you cease all violation of this License, then your license from a particular copyright holder is reinstated (a) provisionally, unless and until the copyright holder explicitly and finally terminates your license, and (b) permanently, if the copyright holder fails to notify you of the violation by some reasonable means prior to 60 days after the cessation.

Moreover, your license from a particular copyright holder is reinstated permanently if the copyright holder notifies you of the violation by some reasonable means, this is the first time you have received notice of violation of this License (for any work) from that copyright holder, and you cure the violation prior to 30 days after your receipt of the notice.

Termination of your rights under this section does not terminate the licenses of parties who have received copies or rights from you under this License. If your rights have been terminated and not permanently reinstated, receipt of a copy of some or all of the same material does not give you any rights to use it.

10. FUTURE REVISIONS OF THIS LICENSE

The Free Software Foundation may publish new, revised versions of the GNU Free Documentation License from time to time. Such new versions will be similar in spirit to the present version, but may differ in detail to address new problems or concerns. See http://www.gnu.org/copyleft/.

Each version of the License is given a distinguishing version number. If the Document specifies that a particular numbered version of this License "or any later version" applies to it, you have the option of following the terms and conditions either of that specified version or of any later version that has been published (not as a draft) by the Free Software Foundation. If the Document does not specify a version number of this License, you may choose any version ever published (not as a draft) by the Free Software Foundation. If the Document specifies that a proxy can decide which future versions of this License can be used, that proxy's public statement of acceptance of a version permanently authorizes you to choose that version for the Document.

11. RELICENSING

"Massive Multiauthor Collaboration Site" (or "MMC Site") means any World Wide Web server that publishes copyrightable works and also provides prominent facilities for anybody to edit those works. A public wiki that anybody can edit is an example of such a server. A "Massive Multiauthor Collaboration" (or "MMC") contained in the site means any set of copyrightable works thus published on the MMC site.

"CC-BY-SA" means the Creative Commons Attribution-Share Alike 3.0 license published by Creative Commons Corporation, a not-for-profit corporation with a principal place of business in San Francisco, California, as well as future copyleft versions of that license published by that same organization.

"Incorporate" means to publish or republish a Document, in whole or in part, as part of another Document.

An MMC is "eligible for relicensing" if it is licensed under this License, and if all works that were first published under this License somewhere other than this MMC, and subsequently incorporated in whole or in part into the MMC, (1) had no cover texts or invariant sections, and (2) were thus incorporated prior to November 1, 2008.

The operator of an MMC Site may republish an MMC contained in the site under CC-BY-SA on the same site at any time before August 1, 2009, provided the MMC is eligible for relicensing.

How to use this License for your documents

To use this License in a document you have written, include a copy of the License in the document and put the following copyright and license notices just after the title page:

> Copyright (c) YEAR YOUR NAME.
>
> Permission is granted to copy, distribute and/or modify this document
>
> under the terms of the GNU Free Documentation License, Version 1.3
>
> or any later version published by the Free Software Foundation;
>
> with no Invariant Sections, no Front-Cover Texts, and no Back-Cover Texts.
>
> A copy of the license is included in the section entitled "GNU
>
> Free Documentation License".

If you have Invariant Sections, Front-Cover Texts and Back-Cover Texts, replace the "with...Texts." line with this:

> with the Invariant Sections being LIST THEIR TITLES, with the
>
> Front-Cover Texts being LIST, and with the Back-Cover Texts being LIST.

If you have Invariant Sections without Cover Texts, or some other combination of the three, merge those two alternatives to suit the situation.

If your document contains nontrivial examples of program code, we recommend releasing these examples in parallel under your choice of free software license, such as the GNU General Public License, to permit their use in free software.

MIX
Papier aus verantwortungsvollen Quellen
Paper from responsible sources
FSC® C105338

Printed by Books on Demand GmbH, Norderstedt / Germany